MW01286142

Automation Airmanship®
Nine Principles for Operating Glass Cockpit Aircraft

Christopher J. Lutat
S. Ryan Swah

S. Ryan Swah

Christopher J. Lutat

New York Chicago San Francisco
Lisbon London Madrid Mexico City
Milan New Delhi San Juan
Seoul Singapore Sydney Toronto

Cataloging-in-Publication Data is on file with the Library of Congress

Automation Airmanship®:
Nine Principles for Operating Glass Cockpit Aircraft

1 2 3 4 5 6 7 8 9 0   QFR/QFR   1 2 0 9 8 7 6 5 4 3

ISBN: 978-0-07-181586-4
MHID: 0-07-181586-4

The pages within this book were printed on acid-free paper.

Sponsoring Editor	Project Manager	Production Supervisor
Larry S. Hager	E. Grishma Fredric, Newgen Knowledge Works Pvt. Ltd.	Pamela A. Pelton
Acquisitions Coordinator		**Composition**
Bridget Thoreson	**Copy Editor**	Newgen Knowledge Works Pvt. Ltd.
	Patti Scott	
Editorial Supervisor		**Art Director, Cover**
David E. Fogarty	**Proofreader**	Jeff Weeks
	MEGAS	

Automation Airmanship is dedicated to the countless pilots,
instructor pilots and aircrew who, for over a century,
have worked tirelessly to smooth the integration of
advanced technology into safe,
reliable flight operations.

About the Authors

Christopher J. Lutat and S. Ryan Swah.

Christopher J. Lutat, **ATP**, has been an airline pilot for a large global airline for more than 17 years and was a search and rescue pilot for the U.S. Coast Guard for eight years. He currently flies the Boeing MD-11 for a major global airline. He is a founding owner of Convergent Performance, LLC, and serves as President of its Government and Aerospace division. Captain Lutat is also a Check Airman and has been an instructor in advanced technology, wide-body global transport aircraft since 2003.

S. Ryan Swah, **ATP**, is a retired U.S. Navy Captain with experience in attack, fighter, and transport aircraft. He currently flies the Boeing MD-11 for a major global airline and has held positions as an FAA Aircraft Program Designee, Line Check Airman, Proficiency Check Airman, Standards Check Airman, Aircraft Technical Pilot, Flight Standards Manager, and Senior Flight Standards Manager. Captain Swah leverages his expertise in glass cockpit, automation checklists, procedures, and flight manuals as the Content Manager of Convergent Performance, LLC.

Contents

Part III   Harmonizing Essential Crew Capabilities with Flight Deck Automation

Preface

As this book was going to press in the fall of 2012, Pierre Sparaco, now-retired Paris Bureau Chief for the industry journal *Aviation Week and Space Technology*, wrote a brief commentary entitled, "Airmanship Anew." This article was largely centered on the report of Qantas flight 32's recovery, by its captain, Richard Champion de Crespigny, after a catastrophic in-flight engine failure. Sparaco's comments echoed those of many in our industry who believe—even as technology has changed the nature of everything from air traffic control to security to flight deck protocols—that there has been too little emphasis on the concept of airmanship compared to the focus on improving profit margins gained from efficiencies and cost cutting. Along with the dramatic engines-out landing on the Hudson River of U.S. Airways flight 1549, Qantas flight 32 joins the ranks of those popularly recognized heroic saves resulting from the display of expert airmanship—and rightfully so. Both of these now widely discussed events seem to evoke qualities of a special kind of airmanship that their crews displayed in saving the lives of hundreds of passengers and crew, following catastrophic failures aboard normally reliable, highly advanced aircraft. It is the search for this "special kind of airmanship" that has led us for the past decade into the field where we have been involved in helping organizations to adopt advanced aircraft of all kinds, and ultimately, to write this book.

In the early part of 2003, as a result of our work with the U.S. Marines and U.S. Coast Guard (both organizations had just begun acquiring aircraft with new, highly automated flight decks), we began to sense that an undertaking like this book might be necessary. The airplane that both organizations were bringing into service was a completely new version of the venerable—*and legendary*—Lockheed-Martin C-130. The new airplane was the C-130J, and both the Marines and the Coast Guard had been flying earlier models of C-130s for decades. We were in the midst of projects with both organizations to streamline procedures and flight deck processes to fit their individual cultures and each organization's unique mission.

In the process of creating everything from new checklists and procedures to training and evaluation standards, we began to come up against a common problem at nearly all levels of the project. Our colleagues in uniform (the Marine and Coast Guard flight crews we were working with) simply wanted to capture the lessons from industry (mostly in civil air transport) of almost two decades of experience with glass cockpit transports. They would blend the experience of the broader industry into their own operations, capitalizing on industry best practices, and thus bring additional safety margins and operational efficiencies to their own flight operations. We wanted to transfer this knowledge to them, and in the process help both organizations through their respective acquisitions without the accidents and incidents that often accompany such leaps in technological capability and complexity. Most of this was fairly straightforward, made easier by our entire team's rich experience in both military and civil transport aviation. Both organizations had unique missions for the C-130J (the Marines for transporting troops and equipment and serving as airborne refuelers for tactical aircraft, the Coast Guard for transport, search and rescue, law enforcement, and humanitarian missions). Both organizations also operated the same "basic aircraft" but with significant differences in aircraft configuration, mission equipment, and even the composition of the "normal crew" that the aircraft would be operated with. And though we were successful in applying many of the industry's early lessons in adopting advanced aircraft, the same problems continued to the surface, no matter what organization we were involved with.

You might think—as we first did—that the problems we encountered would be unique to each organization, and to the unique configuration of each service's aircraft. Yet the challenges we most frequently encountered were basically the same: *how to impart the successful practices of experienced glass cockpit crewmembers from across the industry on an operation with little experience in operating fully integrated glass cockpit aircraft, without requiring years of service gaining experience, and without the attendant lessons of accidents and incidents that are not uncommon in such transitions.* As we have already said, some of this was easy, and we were able to build such practices directly into the procedures, checklists, and other SOPs that these robust flying organizations depend on. The problem, however, persisted and was in every case reduced to the simple fact that the more nuanced skills and techniques of experienced glass cockpit crewmembers staunchly resisted such "proceduralization," and it would be up to each operator to provide the follow-through required to build this airmanship discipline within their flight crews. Although we knew of most of the elements that comprised this family of skills and successful practices, we could do little more than add them up and provide them as a list of recommendations—and then hope they took hold. We went on from these projects to experience the same results with customers in other organizations, civil and military, always with similar outcomes. At the end of a project, we had the same "extra parts" left over, comprised mostly of a list of skills, techniques, and "tricks of the trade" that experienced top performers had been accumulating for decades, which we could not fit into the detailed operational materials we were providing.

By 2005 it was clear that it was time to organize this family of information and bring it more concisely to our colleagues across the industry. And so at the 2005 North American Corporate Aviation Safety Seminar (CASS), sponsored by the Flight Safety Foundation, we rolled out the foundational concepts that would later lead to the principles outlined in this book. Soon after that we began to learn more about other rising disciplines within and outside of aviation that would sharpen our knowledge across every area of glass cockpit airmanship, further refining what we felt had become the core principles of operating complex aircraft in an increasingly complex environment. Over the next several years we continuously worked to simplify and clarify what we consistently found were the central skills of experienced top performers in our field. Eventually these were organized into 12 skills and finally, over the course of several more years, into nine principles that could be learned, further developed, and adopted by individual aviators and broad organizations alike. What we present in this book represents not only years of fieldwork and background research, but also the summation of decades of experience by aviators and their supporting organizations engaged in flight operations with advanced aircraft of every type.

As pilots ourselves with experience in both traditional "round-dial" or "steam-gage" aircraft and advanced global transports, we know that the demands on our own personal airmanship have changed over the past several decades, and we feel quite confident that this is true for many thousands of other aviators as well. As instructors, check airmen, and procedures and training designers, we are quite aware of the demands that technology has placed on cockpit and mission crews worldwide. Likewise, across every organization in our industry, the pressure to preserve increasingly precious assets has never been greater than it is today. Even with all the changes that technology has brought to the industry over the past decades, we are attempting not to redefine airmanship as much as to update its practice wherever automation factors into its execution. We believe that certain aspects of airmanship are timeless and enduring, and that much has been written about those qualities. To each of the authors, *Automation Airmanship* can be defined simply as *the understanding and application of automation to airmanship, to ensure balanced situational and mode awareness and crew workload through the full realm of automation, from no automation to fully coupled, in order to provide for the safest and most efficient flight.* We are not seeking to be "additive" to the fundamental notion of airmanship, but rather to be multipliers of how airmanship is practiced.

Much remains to be learned about glass cockpit airmanship, and we hope that this book helps to generate the enthusiasm and resources to conduct this research and debate across the industry in the coming decades. Along with Mr. Sparaco, we feel strongly that "Principles governing airmanship certainly deserve better treatment and, without reinventing the romance of flying, could produce astonishing results." We are pilots, not human factors experts, flight deck engineers, or aircraft designers. We want what all pilots want— clear objectives to accomplish and a reliable system of actions that support the mission. Although we discuss human factors and complex aircraft and systems in this book, our emphasis is on the essential knowledge from both of these domains that we feel is required for aircrews. We discuss both new and timeless concepts of human factors that influence outcomes on the advanced flight deck, beyond what is traditionally taught in human factors training. At a minimum, we hope to blend a simplified (for pilots) foundational knowledge of the technology and a new way of looking at performance on the flight deck with nine core principles around which individuals can organize their own successful habits. We are convinced that this will result in not just an accident- and incident-free career for many pilots and mission crews, but a career that returns much more professional satisfaction to pilots and aircrews than ever before.

The book is organized in five parts: Part I outlines the foundational knowledge that allows the reader to understand the principles presented in Parts II, III, and IV (the nine principles themselves), followed by Part V that shows how these principles have been successfully integrated into flight operations within several unique and diverse organizations, all of which operate advanced aircraft. Following each chapter that deals specifically with the individual principles, there is a short Checklist for Success that can be applied directly to flight operations of any size and scale. At the end of the book, an appendix contains a definition of each principle and examples of best practices and techniques. Throughout the entire book we provide detailed footnotes that provide the most careful readers with the necessary information to investigate further the research and knowledge we use to build the case for a concise family of principles around which glass cockpit airmanship should form.

We want this book to be both a source of sound technical information and an inspiration for every pilot, instructor, crewmember, support crew, safety team, manager, commander, or even enthusiast to seize upon to help guide a career of understanding and professionalism in flying glass cockpit aircraft. What we have learned ourselves in accumulating the experience and knowledge that went into this book has deepened our appreciation for the amazing aircraft that comprise the global aircraft fleet of the twenty-first century, and greatly contributed to our own satisfaction in pursuing our own love of flying in just a few of those aircraft. Our simple hope is that you will experience the same result.

CHRISTOPHER J. LUTAT AND S. RYAN SWAH
Memphis, Tennessee

Notes

1   Pierre Sparaco, "Airmanship Anew," *Aviation Week and Space Technology,* October 8, 2012, p. 19.
2   Ibid., p. 19.

Acknowledgments

The authors would like to acknowledge the following individuals and organizations that have, over the span of the last decade, worked with us in the field or encouraged us in some other way to complete this book. For making their time available to us for the many questions and feedback on expert performance and facilitating our fieldwork through collaborations within industry and allied air forces around the world, we are grateful and humbly appreciative. Our first collaborative group at Convergent Performance: Lt. Col. Julie Petrina-Curlin USAF, Chris Stickney, Capt. Bob Rayhill, LtCol Kristi Brawley USAF. Our colleagues at Convergent Performance, LLC, especially Dr. Tony Kern, Pat Daily, Dr. Ken Stahl, George Stamper, Greg Rankin, Tom Beck, Ken Zeileck, Will Lutat, Carolyn Goul, Jack Santucci, Bob Novotney, Tom Grazianno, Kerry Reifel, Kacy Speiker-Vorce, Tom Weber, Marcie Miller, Jody Bodeigheimer, Paul Miller, and Christi Janssen. The women and men of the USCG C-130J APO (2004–2007), especially CDR Dan Walsh USCG, CDR Roy Eidem USCG, Capt. John Hardin, CDR John Boris, and Capt. Tom MacDonald. The women and men of the U.S. Marine Corps Air Station Cherry Point, VMGR-252 (2004–2009), especially Maj. Tom Beck, Maj. Jim Palmer USMC, and Msgt. Stacy Crawford USMC. The women and men of the Royal Canadian Air Force, First Canadian Air Division, and office of Air Force Standards Office. The women and men of the Maryland Air National Guard, the 135th Airlift Squadron, especially Lt. Col. Joe Brophy USAF, Lt. Col. Mike Mentges, and Lt. Col. Greg McCleary. Lt. Col. Randy Russell of the USAF ANG. Group Captain Steven Bucholtz, RAAF, Group Captain Kevin Bruce, RAAF and CWO Scotty Campbell, RAAF. Major Scott McKenzie, RNZAF, and Squadron Leader Allan Baker RNZAF. Mark Cross, Sammy Thomas, and Bill Hackathorn of L3. Com. Gary Wallace and Dempsey Solomon of Global Military Aerospace Corporation. Our counterparts at Alenia Aeronautic, especially the C-27J flight test and engineering team. Of the USAF C27J flight publications team, Mr. Scott Humphrey.

At CMC-Esterline Kanata, Ottawa, Canada, especially Dave McKay and Bob Kobierski. Wes Clymer of WBB, Inc., Maj. Gene Mamajek USMC, and Col. Rick Uribe USMC. Dan Shearer of ARINC. Of Bombardier Aerospace, especially the staff of Safety Standdown, Janet Scheibelhutt, Capt. Rick Rowe, Shelley Cox and Kristen Williams. Captain Gene Cernan, USN (ret), for his inspiration and early encouragement. The Royal Aeronautical Society in London, United Kingdom, and the Flight Safety Foundation for allowing us to share the stage with other presenters during the first few presentations of Automation Airmanship to industry, 2005–2008. At McGraw-Hill Professional, the authors wish to thank Larry Hager, Bridget Thoreson, and David Fogarty for their support and patience. Also at McGraw-Hill, Pam Pelton, Patti Scott and Jeff Weeks. At Newgen Knowledge, E. Grishma Fredric.

PART I

The Call,
the Concept,
the Technology

Our theory posits internal mechanisms of great extent and complexity, and endeavors to make contact between them and the visible evidences of problem solving. That is all there is to it.
—Newell and Simon, 1972[1]

CHAPTER 1

The Call for a New Approach to Modern Airmanship

It is handicraft which makes the artist, and it is not in BOOKS that one can learn to manipulate. —Diderot[2]

Airmanship, Error, and Automation: Air Inter 148

On a cold winter evening in January 1992, an Airbus A320 passenger jet operated by Air Inter as ITF 148DA was being flown by its crew from Lyon, near the French border with Switzerland, to Strasbourg, near the Dutch-French border, only about 300 mi away (see Fig. 1-1). The Airbus A320 being operated by the crew of flight 148DA was a derivative of the original A320 and was new to the airline industry in 1992 (the aircraft the crew was flying was barely 4 years old). A highly successful design, the A320 family would become one of the most popular aircraft in the history of commercial aviation over the next 20 years, with more than 3000 in service across dozens of airlines around the world. Since its first flight in 1988, thousands of pilots have, by now, collectively logged millions of hours of accident-free flight time in the A320 family. And as new and updated derivatives of the original begin to enter service today, A320 pilots across the globe will certainly attest to its marvelous handling qualities and the remarkable technology that makes it one of the most enjoyable commercial aircraft to fly today, more than two decades after its introduction. Almost certainly, the crew of Air Inter 148DA counted themselves fortunate to be flying one of the most advanced and capable aircraft in service. Following a short and uneventful flight from Lyon, the crew of flight 148DA prepared for a normal descent and approach to Strasbourg, no doubt looking forward to a routine arrival.

Figure 1-1 The approximate flight track of Air Inter 148DA, January 20, 1992.

On their initial radio contact with French air traffic control nearing Strasbourg, the crew made a normal, predictable request to arrive via an ILS approach to runway 23 (an instrument approach which combines an electronic "glide slope" with a "localizer," to provide precise vertical and lateral guidance to the runway) followed by a visual transition to another runway, runway 05, in order to reduce the time and distance the aircraft would have to cover during the arrival. The ILS approach would allow the crew to descend to a safe altitude above runway 23 and then "break off" the approach in order to fly a visual, circling maneuver to runway 05. This was not an unusual request for this crew, or for this airport, and was certainly within the capability of the A320.

Since there was another aircraft preparing to depart from runway 23 and there was an instrument approach to runway 05, the air traffic controller working the flight cleared flight 148DA for the VOR/DME approach to runway 05—not the ILS 23 that they had requested and planned for. Although the VOR/DME approach (depicted in Fig. 1-2) was not a precision approach like the ILS, it did have the advantage of allowing the flight crew to align the aircraft for runway 05 without the challenge of a circling maneuver, at night, conducted mainly by visual (outside) reference. The approach that air traffic control was directing flight 148DA to fly did not meet the crew's objective of saving time and distance, but it had other advantages. The tradeoff that this crew faced in accepting the VOR/DME approach certainly outweighed the risk involved with the precision, ILS approach to runway 23 followed by a circling maneuver—and is like many

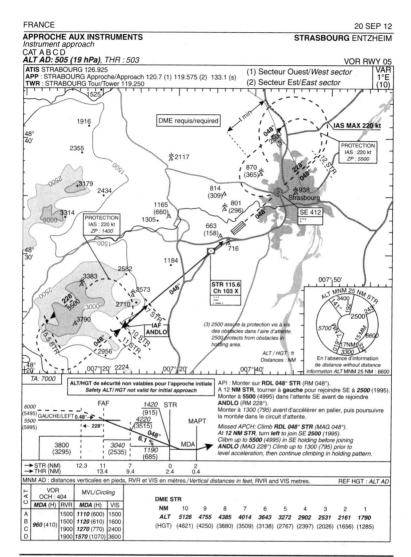

FIGURE 1-2   The VOR RWY 05 approach chart for the Strasbourg-Entzheim airport.

thousands of decisions made daily by experienced flight deck crews around the world.

During their descent, in radio and radar contact with Reims Control (the air traffic control for that region of France), the pilots of flight 148DA continued to discuss the change to their original plans, conduct normal routine checklists, and communicate with the passengers and cabin crew. They accurately recognized that they were entering icing conditions and successfully turned on the aircraft's anti-ice systems, a "routine" distraction that was handled with predictable ease by this crew, but a distraction nonetheless. The change in expected runways and the "work" which that created on their advanced flight deck would contribute to the spike in the crew's workload; but with no other complicating conditions or problems with the aircraft, this, too, was manageable. At the height of their workload for this flight, the crew of flight 148DA was given a radar vector by air traffic control to ANDLO, a waypoint 11 mi on the straight-in, final approach course to runway 05. It would seem that the controller was serving up runway 05 on a platter for this crew—whose aircraft, procedures, and automated systems were designed to execute approaches like the VOR/DME 05 with relative ease. They would have to manage the flight path of the aircraft through a combination of manual and automatic controls with the same high level of knowledge and skill that many cockpit crews must on a routine basis—and that this crew had likely done many times before.

After they reached ANDLO, the approach vertical profile called for a 3.3° glide slope, or angle of descent, to the airfield, for the final descent portion of the approach. Although this was slightly greater than the normal 3° glide slope afforded by most ILS approaches worldwide, 3.3° is both well within the capabilities of the A320 and not unusual for similar, "nonprecision" approaches at hundreds of commercial airports around the world.

The crew of flight 148DA chose to fly the approach to runway 05 "fully coupled" to a point just a few hundred feet above the airport. Again, this is a common practice in which the autopilot and autothrottles are engaged and controlling the aircraft's speed and vertical and lateral flight path with respect to the approach course to the runway. Surely by now, the crew had long forgotten their initial intentions for an ILS to runway 23 and was fully committed and well along in conducting preparations for the VOR/DME approach to runway 05. Their decision to fly the approach fully coupled was sound—a practice that many global operators of advanced technology aircraft have since mandated as "normal procedure" for nonprecision approaches, allowing the flight deck crew to perform the higher-level functions of monitoring the aircraft's flight path while the more routine task of flying the aircraft is delegated to the precise and predictable flight guidance system, autopilot, and autothrottles. In view of the fact that many aviation human factors researchers have long proved—and continue to find new ways to prove again—that the final approach phase of flight often carries with it the highest workload of the entire flight, this was an excellent decision by the crew, allowing them to delegate the physical "flying" to the automation, while monitoring the aircraft performance in preparation for the landing, which would be done without the autopilot.

Acting in accordance with their procedures, the crew programmed the aircraft's flight guidance system to fly the required 3.3° glide path by manually selecting −3.3 on the flight control unit (FCU), which is the panel in the cockpit that allows the pilots to make inputs and modifications to the aircraft's vertical and lateral flight path. Thus, with the FCU electronically coupled to the flight guidance system and autopilot, A320 pilots can—just as all pilots frequently do in the "glass cockpit" of aircraft like the A320—control the aircraft flight path through an electronic interface (literally, "flying" the aircraft with their fingertips through electronic interfaces such as buttons, dials, and knobs), and not the aircraft controls. In approaches such as the VOR/DME runway 05 at Strasbourg, when the pilots judge that the aircraft is in a safe position to land with respect to the runway, the autopilot is disconnected, the pilots land the airplane using conventional flight controls (in the case of the A320, a "sidestick" controller located at both pilots' sides and throttles placed in the center of the cockpit). The pilots of flight 148DA were doing what pilots of automated aircraft frequently do—they were using the automated systems to efficiently manage their workload, allowing the limited processing capacity of each pilot—the *wetware*—to be used for higher-level tasks such as monitoring, communicating with air traffic control, configuring the aircraft for landing, assessing the developing situation with the aircraft's flight path, and the kind of nuanced decision making that cannot be delegated to automated systems.

After a century of powered flight, aircraft accidents and incidents continue to favor the weakest component in the highly dynamic, tightly coupled system typical of high-reliability organizations: the human operators. The crew of flight 148DA, at the moment they programmed the FCU to accept a −3.3° glide path, also were required to do something else—something that has proved to be much more difficult than programming the automation—not just for this crew, but for countless pilots who have since made similar mistakes, some of whom did not live to recall and apply the lesson on future flights. At the critical moment when −3.3° was programmed into the FCU, neither pilot recognized that the autoflight system was in heading/vertical speed (HDG/VS) mode, instead of the correct track/flight path angle (TRK/FPA) mode. By conducting only what equates to just a portion of the complete procedure—*programming*—without *confirming* their actions with the flight guidance and various annunciators in the cockpit, the crew of Air Inter 148DA were commanding the aircraft to fly a vertical speed of 3300 ft/min, almost *four times* the rate of descent that they would certainly have expected the aircraft to fly if they had in fact selected −3.3°, as they had intended. This trajectory, left unnoticed and unmonitored by the crew, would take flight 148DA dangerously short of the runway and leave the aircraft in an energy state, which would require a feat of heroic airmanship to reverse.

One would think especially with the knowledge of how abnormal a 3300 ft/min descent would both feel and look to almost any experienced flight crew (whether they were taking their cues from cockpit instruments or the outside environment, or even the "seat of their pants" so often referred to when talking about flying), that the cues would have been dramatic on the flight deck of flight 148DA. These cues would normally cause both pilots not only considerable alarm, but also the kind of conversation that normally goes along with such unusual situations on the flight deck. In fact, neither pilot appeared to recognize that they were in the wrong flight *mode*, and neither pilot said anything about the 3300 ft/min descent being inappropriate during the instrument approach. Although it is impossible to know for certain, it is not hard to imagine that this silence was not simply a result of inattention, but would have more likely been the result of significant confusion – confusion over why the aircraft was seemingly responding in a way that was not consistent with what both pilots thought should be happening, instead of what actually was happening.

Standard operating procedures, or SOPs, are generally accepted by all professional pilots and crews to be inviolable rules by which to operate; and they are, in every case, implemented by the organization's respective command leadership. Dr. Tony Kern, author of more than a few books on the subject of aviation safety, airmanship, and pilot error, has taught many of us the meaning of SOPs. They are not just regulations to reliably guide a pilot's thoughts and actions, but are a written pact between fellow crewmembers such that none will allow the others to even *accidentally* violate them and in so doing risk the safety of the flight.

Although SOPs for automated aircraft have evolved considerably since the first few years of the global adoption of large, advanced technology aircraft, decades of experience operating transport aircraft of the same basic performance envelope as the A320 have left no uncertainty about the significant deviation from standard of such a high rate of descent during a normally routine, nonprecision approach. What is it that caused this crew to fail to respond to the anomaly that their actions caused, while there was still ample time to intervene with the automation? How is it possible that neither pilot attempted to, or even spoke of the possibility to, modify, or even abandon, the programmed solution that the pilots provided when they initially programmed the FCU? Even as we look retrospectively at the accident at "ground speed 0 and 1.0g" two decades later, it is hard for the typical, inquisitive pilot or knowledgeable observer to reconcile what had to be obvious to the flight crew with their inaction during the exciting final minutes of flight 148DA. Still, the aircraft (which was not equipped with a technology known as *ground proximity warning system* (GPWS), which is now standard aboard most transport aircraft[3]) struck the ground 3 mi after crossing the ANDLO intersection, at an altitude of 2620 ft, near Mt. Sainte-Odile. Among the 96 passengers and crew who had embarked on Air Inter flight 148DA at Lyon, only nine survived.[4]

Certainly, the accident analysis report issued months later by the Bureau d'Enquêtes et d'Analyses (BEA), France's equivalent of the National Transportation Safety Board (NTSB) in the United States, would examine exhaustively the flight data and voice recorders, as well as every conceivable detail associated with the accident. And finally, in 2006, litigation associated with the accident would finally be settled.

New Technology, Traditional Institutions

Since 1992, several other accidents involving advanced technology aircraft made by virtually every manufacturer of commercial aircraft have each in their turn seemed to eclipse the significance of Flight 148DA, leaving the legacy of Flight 148DA obscured by time and since eclipsed by other accidents with similar aircraft. In fact, flight 148DA was not "accident one" among this new type of highly capable, highly complex family of aircraft which now comprises virtually all of every global and national airline fleet, worldwide. Indeed, it was not even the first accident involving the now venerable A320. Flight 148DA simply exists as one example from which pilots, airline training departments and, most recently, military organizations choose to examine and occasionally discuss on the way toward eliminating the risks that a similar accident could ever occur again.

Yet similar accidents do occur with what seems like alarming regularity. During one highly unusual, 5-month stretch from February to June of 2009, three major accidents occurred involving advanced aircraft built by three different manufacturers, operated by three different airlines, all with fatalities.[5] Yet we continue to be shocked and surprised when accidents happen, in part because we have grown to appreciate the reliability of automatic systems across our entire lives, not just in aviation. As pilots and crew, it is considerably more alarming to learn of such accidents on the evening news, especially when considering that aircraft designers, manufacturers, and the marketing departments which serve them have convinced so many in the global aircraft market that these aircraft are virtually foolproof, offering safety margins that can drive the probability of accident or failure to almost zero.

There have been other accidents—many known to all and some known to only a few—from which we might draw a list of virtually the same questions that we have about Air Inter 148DA:

- How could pilots have neglected to complete a routine procedure by failing to conduct an adequate cross-check of the flight guidance system after entering a flight guidance command?

- How is it that the mode selected by the crew did not meet with the mental model of the mode intended by both pilots, yet did not produce counteractions or even a discussion of the incompatibility?

- What caused the crew to give away so much of the ground that pilots have traditionally held to the automation, without any intervention when it was clear the automation was not working as intended?

- How did the pilots fail to properly monitor the aircraft's flight path during what they both certainly knew was a critical phase of flight?

There are many more questions, and almost every accident involving advanced aircraft, large and small, civil and military, commercial or private, leaves investigators and analysts with similar questions, many unanswered. We will examine a few more accidents on the way to the conclusion of this book, as well as some seemingly "heroic" saves that provide dramatic evidence proving that accidents of this kind, involving complex aircraft, are not inevitable.

Aviation has proved to be a remarkably resilient industry in the face of dramatic—*though rare*—accidents that frequently call into question the science of aircraft design and the art of piloting. Like so many accidents prior to flight 148DA, and since, investigations proceed, findings are publicly proclaimed, and lessons are added to the many that already exist. The lessons stemming from accidents in *advanced* aircraft have been accumulating for nearly two decades, which is certainly long enough to generate a unified reaction by pilots and ownership, and even a comprehensive, systematic response that could serve to eliminate the errors that seem so common in so many accidents. As the evidence has piled up as a result of many now well-known failures, the industry has been less focused on collecting evidence from successful outcomes. Yet even when experienced pilots, managers, training designers, and instructors (just a few of the positions held by the authors) search for a unified strategy to combat what is certainly a unified family of likely errors, none can find a singular approach, or one that even comes close or is ready to be put into action.

For decades, rigorous and ever-expanding disciplines have developed around aviation safety: aviation human factors, accident investigation and reporting, flight simulation and training, as well as aviation psychology. These disciplines, and others, form a veritable patchwork of countermeasures that have been stitched into a wide variety of formats and media for the consumption of professional pilots and businesses, whose existence increasingly depends on safely operating glass cockpit aircraft. There are so many countermeasures that dedicated aviation professionals have hundreds of annual professional gatherings and a wide variety of professional journals to choose from where they can hear about or read the latest research results and absorb the most exciting new developments in flight safety and human factors. Crew resource management (CRM), risk management, threat and error management (TEM), to name just a few safety initiatives, have all run their course and permeated the industry so thoroughly that it would seem that all has been said, printed, or discussed in one of the hundreds of panel discussions that have taken place before and since Air Inter flight 148DA flew into Mt. Sainte-Odile in January 1992. Search as we may, we still cannot place our hands on a singular, widely accessible, comprehensive approach to the known problems, and known solutions, to safely operating advanced technology aircraft.

Since the dawn of aviation, the technology has come in advance of the operators. Writing about the advances in technology that were occurring almost 80 years ago, Franklyn E. Dailey, Jr., accurately observed, "Instruction follows innovation in performance; it does not lead. Pilots came before instructor pilots."[6] At roughly the same time that the industry was absorbing the shock of accidents involving what were surely the most reliable and safe aircraft ever built (at about the time that flight 148DA tragically impacted short of runway 05 in Strasbourg), practical pilots and flight instructors, and other professionals who lived with and operated these systems every day, began an often informal process that would ultimately lead to the many components of *Automation Airmanship*. Pilots—not the designers, researchers, and engineers—were learning to cope with the new technology and adopting strategies, habits, and routines that they were constantly shaping into practical tools for themselves, for their colleagues, and, if they were instructors and evaluators, for their student pilots. Nearly every experienced pilot of a glass cockpit aircraft can recall a "tip" or "trick" that he or she routinely uses in the practice of airmanship on the flight deck today, that was learned from some experienced instructor or check airman, or the revered and romanticized "salty old airline captain" (they *do* exist). The very best and most experienced instructors and captains are known, and admired for their ability to concentrate a large amount of knowledge into just a few useful and memorable "aids" which help to simplify the complex flight deck and which help crews organize and prioritize their flight deck actions. We all know a few great instructors, check airmen, and captains who have provided us with that one piece of knowledge that, like some brilliant insight, made simple so much complexity.

Unfortunately, much of this information is passed around the profession informally, outside of the research community and safety industry that support modern aviation. This is not surprising, given that for almost a century flight training has always had some informal quality about it that makes both the instructor, or veteran, and the student, or novice, feel as if they are passing on or receiving some special, unique knowledge or technique that, if internalized, can become another nugget of durable knowledge for the successful pilot.

Thus, seemingly invaluable skills in flying highly automated aircraft are often passed along like tribal wisdom: "You won't find this written down anywhere . . ." or "We do it this way on the line; forget what they taught you in the schoolhouse . . ." or even this one: "The most important thing you can do every day isn't part of the formal procedure." Many of these fireside lessons are invaluable, and all have come at the price of someone else's (sometimes fatal) mistake.

For nearly two decades, insightful veteran pilots, and many instructors, have been piling up these useful trade secrets, without so much as a Ph.D. in human factors or a Master's degree in instructional systems design. They have watched expertise develop, and they pass on as much as they can communicate to their less-experienced colleagues. Although accidents have produced their own truths, and researchers have generated valuable data and insightful findings across so many facets of the profession, we finally must recognize that collectively, though somewhat informally, glass cockpit airmanship has shaped itself into a largely intact body of expertise whose time has come for a full synthesis and open understanding across every organization that comprises our profession, and with all pilots who aspire to, or now find themselves in the cockpit, of today's advanced aircraft.

What we have come to perceive as the most effective flight deck practices have come from not only the most admired instructors and veteran captains. During the decade after the Air Inter accident in Strasbourg, some far-thinking researchers, both within and outside of aviation, were formulating valuable new approaches to expertise and expert performance, and gathering findings that would provide significant, complementary knowledge surrounding the discipline of the rapidly developing environment of glass cockpit airmanship. These individuals were bringing to maturity such disciplines as cognitive task analysis (CTA), naturalistic decision making (NDM), and advanced qualification program (AQP) training, among dozens of others. In the 1990s we also began to see that much was being learned about the way the human brain works, and how the brain learns new information and formulates decisions from an increasingly complex and high-risk environment. Scientists such as Robert Hoffman, Gary Klein, and Joseph LeDoux were all making significant strides in related fields that have helped significantly in developing a unified approach to making decisions in complex, tightly coupled systems that operate through complicated and often automated control channels, in high-stress/high-risk situations. The field of *human-computer interaction*, or HCI, has been maturing so rapidly during the past 20 years, and providing so much valuable insight on how humans interact with computers and automation, that it is difficult to reduce the information to a manageable amount of practical knowledge for aviation. Thousands of experts on safety and accident analysis working for government-sponsored organizations such as NASA and the FAA, at private institutions and foundations, international organizations, and private airlines, have provided invaluable understanding to the role of the human component in the complex system of contemporary aviation.

From Chaos and Complexity: Order and Excellence

For those involved in any dimension of the safety industry, particularly the aviation safety industry, it seems supremely difficult to proffer a comprehensive solution to a broad family of problems associated with such a complex, tightly coupled system as operating advanced aircraft in today's increasingly complex global aviation environment. There are so many components to this system that the complexity is staggering: not only complicated human factors involving those associated with skilled piloting, but also human factors engineering, ergonomics, and the myriad complexities involved in designing reliable interfaces between the highly trained crew and the aircraft itself—a tremendous feat of industry encompassing the latest in materials science, computer software design, propulsion, and flight controls (both power-by-wire and fly-by-wire systems are now the norm in most modern as aircraft) as well as dozens of other complicated support systems. Surely any comprehensive strategy to counter the risk of errors in such a complex system must involve years, if not decades, of research and hundreds of thousands of pages of analysis before the data are turned over to the designers of procedures, training, and evaluation programs for the hundreds of thousands of pilots and crews across the industry.

Or does it?

Fueled by the many seemingly parallel fields of research, for the past decade we have been combining a variety of disciplines that have been converging—intentionally and otherwise, but converging nonetheless—to provide a wide range of services to organizations as small as one-aircraft business jet operators, to large Western defense forces and their component branches, simplifying the transition from traditional or "legacy" fleets to the modern, up-to-date advanced aircraft fleets that dominate the industry today. This work has included task analysis, courseware design, interface and procedures development, and flight crew training and evaluation. Because of the need to make sense of the technology and define the crew experience for those new to advanced aircraft, we have employed a variety of disciplines and methods to concisely define expert flight deck performance for glass cockpit crewmembers (Chap. 13 discusses this in depth). Experience in designing procedures and flight crew interfaces helped us to pinpoint the essential flight crew skills that resist being turned into procedures and are included in many assessments of expert performance in high-risk/high-reliability domains such as aviation (Chap. 14 discusses procedures design in depth).

While all this has formed the core of our collaborative work involving advanced aircraft, we have also maintained active professional pilot careers, operating some of the world's largest and most advanced and capable transport aircraft while serving as airline captains, flight standards managers, check airmen, and instructors. Throughout all this work and in combination with real line experience, we have come to appreciate the many contributions of a wide range of professionals, and we believe that our consolidated views on a composite approach to pursuing a safe and rewarding career in flying glass cockpit aircraft is now inescapably within the reach of every pilot—experienced and novice, civil and military, young and old. We propose to provide the foundational principles of this discipline in the following chapters of this book.

The contemporary business author and speaker Marcus Buckingham has researched the phenomenon of excellent performance and writes extensively about it in his 2005 book, *The One Thing You Need to Know*. Buckingham explains the concept of the "one thing" by first asking, "Why are some explanations more powerful than others?" Instead of attributing them to deep truths, he suggests that *controlling insights* are at the center. More compelling, however, is what Buckingham asserts comprises a controlling insight—the building blocks of the "one thing" you need to know about almost any discipline. The test, he asserts, for something to qualify as the one thing is threefold: first, it must apply across a wide range of situations; second, it must serve as a multiplier; and third, it must guide action: "In short, no matter what the subject, the controlling insight should not merely get you onto the field of play. It should show you how to win and keep winning the game."[7]

The universal nature of the test for the "one thing" certainly helps to explain how a few glass cockpit pilots have come to help so many others develop their own expertise much more rapidly than their predecessors. It also helps explain how findings within aviation, and in related disciplines, can contribute so much to our understanding of how the best pilots interact with automation. In short, it helps us to capture these "controlling insights"—which to pilots equate to the best skills and techniques—consolidate them, and present them in a systematic way so that many can profit from the supreme efforts and, in some cases, *supreme sacrifices* of the few.

In the following chapters we attempt to represent the best of all the research, the accident and incident lessons, and the most practical experience collected from interviews and observation of some of the finest and most experienced pilots, instructors, and check airmen in the business, as well as the myriad researchers, of whom a few have already been mentioned. We will present the basic environment that all glass cockpit pilots find themselves in by introducing the very technology that comprises the modern flight deck, objectively and without preference to any manufacturer. We will put the *contemporary* in the context of *history*, so that every glass cockpit aviator will better understand the foundations on which the current technology was built over 75 years ago and still relies as we surge into the future of air transport. In subsequent chapters we will introduce the *Nine Principles of Automation Airmanship*—designed to be applied by any pilot of any advanced aircraft (and useful for pilots of less advanced aircraft, as well)—along with other concepts, which will undoubtedly help all pilots manage their equipment and environment to higher levels of not just safety and efficiency, but reward and satisfaction in the practice of their own craft. Along the way, we will present evidence from accidents, incidents, and the research community to solidify your individual understanding of what it takes to develop a highly reliable, personal level of Automation Airmanship.

Given all this information, every pilot, instructor, manager, safety professional, or flight crewmember whose performance is somehow connected to the actions of the flight crew in advanced aircraft will come away from this book with a substantially improved understanding of glass cockpit airmanship. What we have known and promoted as a "lens" through which to view airmanship as a whole—*Automation Airmanship*—will become as natural in executing flight deck responsibilities as planning and executing a routine instrument approach. In fact, it will be *the same* as planning and executing a routine instrument approach.

We do not propose to have all the answers, but think that by the end of this book you will surmise that we have put forth a discipline of airmanship, embodied in practical strategies, that meets most of the challenges faced by pilots and crews of most advanced aircraft. If you are an experienced captain, you may see some familiar principles that you have been practicing for many years; only now you will be able to place them in the context of a concise discipline for what may be the first time in your career. As a copilot or first officer, mastering Automation Airmanship with novice credentials may not help you make captain or aircraft commander any faster, but it will help make you a better captain when you do; in the meantime the flight deck crew that you are a part of will be better because of the way you undertake your flying job. If you are part of an organization that operates advanced aircraft as part of the mission crew, are in a command leadership position, or serve in some other specialized support role, you will better understand the dynamics of the glass cockpit airmanship, increasing the level of integration between you and the crew in the cockpit. We will not explain everything you need to know about the specific advanced aircraft that you are checked out in, but we will explain how any individual crewmember can systematically build a reserve of essential knowledge that she or he can rely on across an entire career, no matter the aircraft. We boldly propose that when you have completed this book and begin to apply its actionable steps in your routines as you interact with the technology on the modern flight deck, you will achieve a noticeably increased level of safety and will enjoy greater professional satisfaction than you thought possible on the highly automated flight deck.

Notes and References

1. Allen Newell and Herbert A. Simon: *Human Problem Solving*. Englewood Cliffs, NJ: 1972. Prentice Hall. pp. 9–10.
2. Diderot, D., & d'Alembert, J. (Eds.). (1751–1772). *Encyclopédie ou dictionnaire raisonné des sciences, des arts et des métiers* [Encyclopedia or systematic dictionary of the sciences, arts and crafts]. Paris: Briasson, David, Le Breton, Durand. Page xl.
3. GPWS technology was not installed on the Air Inter A320 fleet because of a number of operational and commercial constraints identified by investigators; in response to the Bureau d'Enquêtes et d'Analyses (BEA) accident investigation, Air Inter equipped its entire A320 fleet with GPWS: "The analysis in §22.36 does not conclude that the carriage of GPWS could have prevented this accident. It does show that statistically the carriage of this equipment is beneficial and, with appropriate procedures, it is very likely that the crew would have reacted positively to the alarm." The French personnel investigating the accident would go on to recommend installation of GPWS on aircraft with capacity of 31 or more passengers.
4. Details were obtained from the official report of the Bureau d'Enquêtes et d'Analyses pour la sécurité de l'aviation civile (BEA): RAPPORT de la commission d'enquête sur l'accident survenu le 20 janvier 1992 près du Mont Sainte-Odile (Bas Rhin) à l'Airbus A320 immatriculé F-GGED exploité par la compagnie Air Inter. [Office of Investigations and Analysis for Civil Aviation Safety (BEA): Report of the Commission of Inquiry into the accident January 20, 1992 near Mont Sainte-Odile (Bas Rhin) with Airbus A320 registered F-GGED operated by Air Inter.]
5. Colgan Air flight 3407 (Bombardier Q400; February 12, 2009: Buffalo, NY, USA), Turkish Airways Flight 1951 (Boeing 737-800; February 25, 2009: Schipol, Netherlands), and Air France 447 (Airbus A330; June 1, 2009: North Atlantic Ocean).
6. Franklyn E. Dailey, Jr., *The Triumph of Instrument Flight*, Dailey International Publishers, Wilbraham, Mass., 2004, p. 56.
7. Marcus Buckingham, *The One Thing You Need to Know*, Free Press, a Division of Simon and Schuster, New York, NY. 2005, p. 15.

CHAPTER 2

Expert Performance on the Twenty-First-Century Flight Deck

Experts have all the time in the world.
—Sir Frederic C. Bartlett, 1958[1]

Applying the Discipline of Expertise

In the early stages of research and fieldwork that led up to the development of Automation Airmanship as a flight deck discipline, we shared one common observation in our experience with superior performance on the glass cockpit flight deck: *We knew superior performance when we saw it.* This theme strikes a chord among many professionals across hundreds of domains requiring high levels of expertise. The challenge we faced was to distill our observations and experience in search of, at a minimum, an *approximate* measure of expertise that could be applied across the vast family of aircraft: including both highly advanced aircraft, and those with only modest levels of automation. It was something that had vexed many of us for years, and it was something that our customers in industry were eager to take advantage of—that formula for achieving success that seemed only achievable in rare and inopportune moments. This, therefore, is where we began our search for the hallmarks of expert Automation Airmanship.

For nearly a century the aviation industry has been developing performance standards for pilots and aircrew and for the crews who build, maintain, and repair aircraft. Regulatory agencies sponsored by governments, international organizations, and countless academic institutions have combined to create a gauntlet of physical and mental tests that aviation professionals must navigate to obtain—and maintain over their careers—the certificates that allow them to practice in their chosen profession. Following World War II and the rapid development of commercial jet air transport in the second half of the twentieth century, this approach produced generally favorable results. And with the introduction of advanced aircraft in the last two decades, these practices were directly adapted to training crews in operating the latest aircraft. At the time, in the struggle to meet the promise of competitive advantage, this approach was both cost effective and logical; in short, it made sense that a system of evaluation that was tested over decades and was under constant scrutiny would serve the industry for the future, just as it had in the past.

Accidents involving advanced aircraft would at first startle the industry and then force modest adaptations to existing training and evaluation schemes. Air Inter 148 (Strasbourg, 1992), China Air 140 (Nagoya, 1994), American 965 (Cali, 1995), and Air France 447 (Atlantic Ocean, 2009) stand out among hundreds of other accidents involving advanced technology aircraft of various levels of complexity. In each case, extensive investments were made in both the hardware and wetware[2] in order to avoid such disasters in the future. Each accident served to shock the industry with news that aircraft that were designed to optimize safety and provide unprecedented levels of reliability could be involved in disasters whose causes could be directly linked to the crews who operated them. In every case, in the face of such startling news, training departments absorbed what lessons could be abstracted from the accident investigations and plodded on in largely the same form as they had been before the introduction of glass cockpits to fleets across the industry.

Unquestionably, valuable insights have influenced the way aircraft are operated as a result of early trials and failures realized through accidents. These accidents and others like them have inspired a generation of investigators, researchers, safety organizations, and the like, which have spawned numerous interventions designed to prevent future accidents. Undoubtedly, the research organizations have made enormous contributions to our collective understanding of the role of technology on the flight deck and of the relationship between the technology and the humans who interact with it. You will find that much of the foundational knowledge of Automation Airmanship is built upon the best research and experience of these valuable institutions. However, unlike accident investigators, researchers, and human factors experts, pilots must cast a wide net to call upon the best that these other communities provide, without the luxury of waiting for peer-review panels to digest the latest advances in human factors science. It is the same with Automation Airmanship—we have relied on interdisciplinary research, accident analysis, research into human-machine interface issues, and, most of all, decades of experience in flying glass cockpits, observing crews of various levels of expertise, in both military and civil settings. Where we think it is important, we state that current research is limited, out of date, or altogether lacking; and we encourage research in these areas. In all areas, however, our aim has been to extract the very best of available insights and make them accessible to all pilots, in every type of aircraft, no matter where on the globe they fly.

In the same way that we know expert performance when we see it, we also know poor performance when we see it. Sir Frederic Bartlett observed over half a century ago what expert performance seems like to the novice: watching an expert work, to the inexperienced, is watching someone perform complex tasks *as if he or she has all the time in the world*. This phenomenon is observable in virtually any activity that requires skill, from sports to brain surgery, from music to flying advanced aircraft of all types. What novices do not see readily is that experts, while performing difficult tasks with ease, are simultaneously reflecting on years of experience and "tacit knowledge." Research psychologist Gary Klein describes tacit knowledge as being "like the part of an iceberg that is below sea level. We don't notice it, and we can't easily describe it. Therefore, we are usually oblivious to the tacit knowledge we use in applying our explicit knowledge of facts, rules and procedures. Expertise depends heavily on tacit knowledge, as do many of our everyday skills."[3] Academics who have studied expertise as a career observe that "There is an element of unencumbered elegance in expert performance."[4] Right now, if you have carefully read the previous sentence, in your own mind you are comparing that statement to your own experience and undoubtedly recounting instances that match the experts' conclusions.

Toward an Emphasis of the Best

There is an enormous volume of work emerging in the burgeoning field of expert performance, and that work has helped to give structure to the principles that we have selected to map in real terms as the discipline of Automation Airmanship. Our focus has been to select the principles that are most easily understood and able to be taught, observed, and fed back to aircrews in order to develop a high level of expertise, as rapidly as possible, while at the same time providing a firm structure upon which experience can be developed. Our emphasis is not, therefore, on what experienced crews are doing at the height of the breakdown of performance, which often results in accidents. (Even so, we suggest that research in this area might be useful.) Although it is important in understanding accidents and their causes, we are most interested in what the *very best* crews are doing when things are going smoothly—the actions that result in *unencumbered elegance.* Here we believe that more research would *dramatically* improve our understanding of what the best crews are doing during successful outcomes, dramatically improving safety. Scholars in the field of expert performance refer to this as "studying minds when they are engaged in successful accomplishment—when things 'work.' "[5] This is also what we are interested in—and what we hope to be able to transfer to pilots and aircrew everywhere through the principles introduced in this book.

Naturally, codifying the best behaviors into one discipline seems to be daunting at least and an elusive goal at best—but one which we will show has been brought closer to realization by invoking some tools used by research across a variety of fields, including aviation but also other high-risk/high-reliability domains.

The cockpit crew of any modern aircraft reflects a combination of years of training and experience of the crew, and the technology itself, being used in cooperation with other complex systems outside the cockpit. It is a daunting mission to gather the best principles at work in this complex equation—and some would say impossible, without years of analysis and expensive research budgets. Perhaps this is so. But then again, if we wait for the research community to lead the way through the performance challenges faced by pilots and crews, we may not achieve, in this generation, the promise that we all sense is possible by matching expert performance to the reliable systems rolling out of assembly hangars around the world. In the face of the many critics of aviation safety and the inexorable creep of automation and technology toward ground that pilots have held for over a century, we cannot sustain an accident rate that routinely compromises public confidence and wastes precious public and private resources.

The same approach has helped professionals in many other industries to codify the actions and indications which can lead experienced professionals to act to change outcomes where novices fail: firefighters choose the right approach to large structural fires, nurses act to save the lives of sick infants, weather forecasters help predict severe weather accurately, military field commanders now make better decisions about risk in combat, to cite just a few examples.[6]

In the case of each of the nine principles of Automation Airmanship outlined in this book, we have looked at not just what expert crews do to produce outstanding results but at *how* they do what they do and how they think about what they do. We have been involved in the practical application of glass cockpit operations as well as cognitive task analysis; we have been instructors and evaluators—professional observers of expert performance—and have been active on the very edge of technology adoption with organizations of all sizes. Most of all, we are pilots ourselves, professionally interested in improving our own performance and that of those around us as we exercise our qualifications in real-world settings. Everywhere we have worked, with customers all over the globe, we have found pilots eager to put into practice these principles, and to reshape the environments in which they work so that these actions can be improved upon and replicated, systematically, over and over.

Gary Klein, one of the most prominent researchers in the field of expert performance and decision making in high-risk organizations, has provided us with a large measure of confidence as we have developed our concepts, enabling us to communicate the most important aspects of glass cockpit airmanship to the broader population of aircrews. He has described the achievement of professional expertise in the way that it relates to our vocation more eloquently than we can ourselves:

> When a person attains a high level of proficiency, we expect to see certain characteristics of performance. We expect the person to be able to make judgments and discriminations that are difficult for most other people. The expert must be able to apply the experience to a wide range of tasks encountered in the domain, including nonroutine cases that would stymie people who are merely competent. The best experts will set the standards of ideal performance for a domain. Finally, experts don't simply know more. They know differently. The breadth and depth of their knowledge allows them to "see the invisible" and to perceive what is missing in a situation, along with what is present.[7]

On January 15, 2009, in a feat of airmanship that has been hailed repeatedly and replayed countless times, Captain Chesley Sullenberger and First Officer Jeffrey Skiles managed to make a safe water landing of a crippled Airbus 320 on the Hudson River in New York City. In the interviews and testimony that took place in the following months, both pilots made clear that they had performed according to procedures learned and practiced in training, whether at US Airways or in their previous military careers. Clearly, their remarkable feat appears to be nothing less than "a miracle" to the untrained and novice, and most certainly to those outside of professional aviation. However, their message that they were performing as they had been trained (a tall dose of humility to most observers) resonates with almost every professional pilot. Their accomplishment, to the experienced, speaks of the value of professionalism and training. Read Klein's description of "a high level of proficiency" above, in the context of their dramatic glide into the river. This interpretation brings to life Klein's words through an unconventional lens—a lens that, for the most part, has been possessed by only a privileged and experienced few who deeply understand the qualities of expertise and expert performance.

Learning Mastery from All Experts

When we confront large and diverse audiences of pilots and aircrews, we often lead our discussion of Automation Airmanship principles with an example of expertise drawn from outside of aviation. We illustrate the concept of the nine principles as an index of expertise for the broader discipline, Automation Airmanship, by drawing on the experience of tournament chess, and mastery, in the domain of a game that most of us are familiar with. Chess is a game that can be enjoyed by all ages (the basics can be learned easily by 4- and 5-year-old children), at a wide variety of skill levels, from novice (recreational players) to intermediate and advanced (club players) to acclaimed masters (tournament players). It makes sense that the science of expertise found much to exploit in the domain of chess mastery—a fixed field of action (the board, comprising 64 squares arranged in an 8-by-8 grid), common rules (each player controls 16 pieces with distinct rules as to their movement), and a very large population to examine, with tournaments being held virtually nonstop, all over the world. Neither of the authors is a tournament chess player, but we both acknowledge that the rankings of chess masters is accomplished through objective standards that express the master's level of expertise, in some ways similar to the progression of a pilot's progress through early interest in flying to the highest level of professional achievement.

A variety of rating systems exist in the sport of chess, and there exist several activities which were devised as an "index of expertise"—a means to identify the truly expert at the game. One of these tasks requires the player to move a knight piece across an otherwise empty board, using legal knight moves, visiting each square only once—a puzzle that dates as far back as the ninth century CE. The time that it takes to accomplish the "knight's tour" has been adapted as one index of expertise in chess. Like many games of strategy, chess can be learned systematically, and expertise can build dramatically with the correct coaching, practice, and experience. Figure 2-1 shows several solutions to the knight's tour—clearly

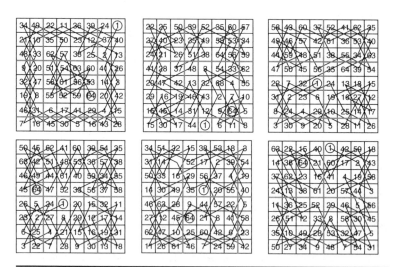

Figure 2-1 Some possible solutions for the knight's tour.

devised by experts. For those of us who operate routinely in the complex environment of the highly automated aircraft, the parallels are obvious: As in chess, so in aviation, there are experts whose skills surpass others' skills. In aviation, however, we are all committed to the outcome where everyone is a winner, where every occasion for crews to succeed is realized.

Translating the concept of the knight's tour to operating a glass cockpit, advanced aircraft is not much of a stretch when you consider what it is that ties together experts across all domains. Klein and the researchers whom he has worked with have gathered a list of key cognitive elements that separate expert performers from novices, across a number of high-risk/high-reliability domains. Whether you are an aspiring chess player, a medical resident starting a rotation in a trauma center, or a crewmember entering the first stages of training on a glass cockpit aircraft, Klein's survey is invaluable in setting personal performance standards for a lifetime of professional achievement. Likewise, for intermediate, advanced, and expert performers, the survey can organize years, or decades, of experience, enabling the performer to access his or her expertise more effectively. Our own research into the discrete actions performed by the highest-performing crews on the glass cockpit flight deck reflects many of the elements found in Klein's survey, and these have been combined in many ways across the entire discipline of Automation Airmanship, including texts, standard operating procedures, checklists, quick reference handbooks for the flight deck, and actual flight manuals, as well as training and evaluation schemes. In Chaps. 4 through 12 we introduce each of the nine principles of Automation Airmanship, allowing us to merge findings like those below with skills that we have been observing in the field and in training and evaluation settings for over a decade. The translation of these concepts from other high-risk/high-reliability domains is direct, and the foundational elements outlined by Klein are outstanding organizational references.[8]

> *Mental models:* Experts have richer mental models than novices or even proficient performers—they understand a wider range of causal connections that govern how things work and can apply them fluidly and flexibly as events change.
>
> *Perceptual skills*: Experts have developed perceptual skills that enable them to notice subtle cues and patterns and to make fine discriminations that may be invisible to others.
>
> *Sense of typicality*: Experts have accumulated patterns and experiences into prototypes so that they can judge when they are dealing with a typical event and when they are facing something that is not quite right and needs attention.
>
> *Routine:* Experts have learned a varied set of routines, so they can usually find some way of approaching problems. Usually, SMEs

(subject matter experts) can just plug a well-learned routine into action, but sometimes the SMEs need to alter a routine or cobble together parts of several routines. In each case, SMEs can use their broad repertoire of routines to adapt to problems.

Declarative knowledge: Experts have a lot of declarative knowledge—lots of factual information, rules, and procedures—that they can draw on.

It is easy to see the immediate application to the observation of expert performance in any domain; but finding this research at the same time as we were formulating our qualities of expertise in the glass cockpit was a powerful motivator in both completing our work and making it accessible to industry. Other research continues to develop rapidly, making this an exciting time to be part of the emerging discipline of codifying expertise. Michelene T. H. Chi, a professor at Arizona State University, provides a characterization of expertise that complements and enhances Klein's survey. Her list, a "characterization of expertise," covers seven major ways in which experts excel:[9]

Generating the best: Experts excel in generating the best solution, such as the best move in chess, even under time constraints, or the best solution in solving problems, or the best design in a design task. Moreover, they can do this faster and more accurately than nonexperts.

Detection and recognition: Experts can detect and see features that novices cannot. For example, they can see patterns and cue configurations in X-ray films that novices cannot. They can also perceive the "deep structure" of a problem or situation.

Qualitative analyses: Experts spend a relatively great deal of time analyzing a problem qualitatively, developing a problem representation by adding many domain-specific and general constraints to the problems in their domains of expertise.

Monitoring: Experts have more accurate self-monitoring skills in terms of their ability to detect errors and the status of their own comprehension.

Strategies: Experts are more successful at choosing the appropriate strategies to use than are novices. Not only will experts know which strategy or procedure is better for a situation, but also they are more likely than novices to use strategies that have more frequently proved to be effective.

Opportunistic: Experts are more opportunistic than novices; experts make use of whatever sources of information are available while solving problems and also exhibit greater opportunism in using resources.

Cognitive effort: Experts can retrieve relevant domain knowledge and strategies with minimal cognitive effort. They can also execute their skills with greater automaticity and are able to exert greater cognitive control over those aspects of performance where control is desirable.

We have come a long way since 1958, when Sir Frederic Bartlett observed that experts at work seem to have all the time in the world. Certainly, the crew of US Airways 1549 had no time to spare in their successful ditching in the Hudson, yet the news media made it seem as if they performed with relative ease. In his 2010 release *Fly by Wire*, author William Langewiesche dramatizes the short 5-minute US Airways flight 1549, describing Captain Sullenberger's performance in words fit for any student of expertise:[10]

> [Sullenberger] . . . had become a diligent pilot. The career had certainly narrowed his experience in life. But he nonetheless . . . was capable of intense mental focus and exceptional self-control. Normally these traits do not much matter for airline pilots, because teamwork and cockpit routines serve well enough.[11] But they had emerged in full force during the glide to the Hudson, during which Sullenberger had ruthlessly shed distraction, including his own fear of death. He had pared down his task to making the right decision about where to land, and had followed through with a high-stakes flying job. His performance was a work of extraordinary concentration, which the public misread as coolness under fire.

This book is not about preparing novices for the lifelong task of developing the judgment and experience embodied by the crew of the "miracle on the Hudson." It is not another treatise on crew resource management, or even threat and error management. It is about developing the elements of expert performance across several principles of airmanship that will result in specific actions that every pilot can execute, during routine, mundane flights in which nothing unusual occurs, and during the ultimate tests that demand our finest performance.

We now turn to the technology itself, giving our readers a sound and reliable framework for understanding the environment of the glass cockpit. Chapter 3 distills the technology down to the bare essentials that all glass cockpit pilots should be familiar with. And putting that knowledge into the context of the historical progression from previous generations of flight deck design will undoubtedly fix in your mind the fundamental simplicity behind all glass cockpit designs. Upon that knowledge base, you will be prepared to build a personal index of expert performance that you can adjust to fit whatever advanced aircraft you fly.

Notes and References

1. Paul J. Feltovich, Michael J. Prietula, & K. Anders Ericsson: Studies of Expertise from Psychological Perspectives. In *The Cambridge Handbook of Expertise and Expert Performance*. Cambridge University Press. New York, NY. 2006. P. 55.
2. *Wetware* is a general term used interchangeably in this book with *human operator* as a partner in the human-machine interface. Some definitions of wetware are limited to describing the brain and its logical and computational capacity. We use the term in the holistic sense to include the brain and the entire physical system that supports human activity on the flight deck.
3. Gary Klein, *Streetlights and Shadows: Searching for the Keys to Adaptive Decision Making*, MIT Press, Cambridge, Mass., 2011, p. 35.
4. K. Anders Ericson (ed.), *The Cambridge Handbook of Expertise and Expert Performance*, Cambridge University Press, Cambridge, England, 2006, p. 55.
5. Beth Crandall, Gary Klein, and Robert R. Hoffman, *Working Minds*. MIT Press, Cambridge, Mass., 2006, p. vii.
6. Ibid., p. 4.
7. Ibid., p. 134.
8. Ibid., p. 135.
9. Michelene T. H. Chi, "Two Approaches to the Study of Experts' Characteristics," *The Cambridge Handbook of Expertise and Expert Performance,* Cambridge University Press, Cambridge, England, 2006, pp. 23–24.
10. William Langewiesche, *Fly by Wire: The Geese, the Glide, the Miracle on the Hudson*, Farrar, Strauss and Giroux, New York, 2009, p. 10.
11. We will show in subsequent chapters that expert performance on the flight deck goes well beyond mere teamwork and routines, as Langewiesche purports.

CHAPTER 3

Fundamentals of Modern Aircraft Automation (for Pilots)

The gyro device that the Sperry family reduced so successfully to practice was a step or two ahead of aviation's ability to take immediate advantage of what that gyro could provide. Aviation had to negotiate the steps in between, and eventually came to understand that essential pilot information was bundled inside the new gyroscopic stabilizer.
—Franklyn E. Dailey, Jr., 2004[1]

The Roots of the Modern Flight Deck

The function of a *flight management system* (FMS) that most pilots would likely hail as its greatest achievement is the ability of the device, through manipulation of the input pathways through the FMS keypad or other input device (multifunction control display unit, trackball, cursor control device, touch screen input, the flight guidance control panel, or other methods), to create a visual representation of the lateral and vertical flight path of the aircraft. This is most commonly represented in modern aircraft as a *magenta line*—a visual line connecting waypoints, navigation aids, runways, and airports along a predetermined route of flight for the crew to see and monitor independent of needles, radials, and Morse code identifiers. These technologies (beacons and navigation aids commonly represented by instruments in the cockpit with needles and compass bearings) have been used by flight crews for almost a century to perform the same basic function: *to visually locate, with certainty, the aircraft's position in space and time,* for the flight crew. The FMS integrates all these earlier technologies, along with other much more accurate and reliable position information (such as GPS) alongside other systems (platform data, air data, etc.) to create the most prominent and relied upon feature of the glass cockpit—the magenta line, which for many pilots is the crowning achievement of the FMS.

More than 80 years ago, an emerging technology had a similar, and arguably more transforming, effect on aviation; understanding that transformation gives foundational meaning to both the contemporary and emerging technologies on today's flight deck. This chapter is for both experienced and aspiring glass cockpit aviators alike; veterans who skip ahead to other portions of this text will be sacrificing an understanding of the role of one of the most fundamental components of the advanced cockpit. Execution of Automation Airmanship at the highest level will be more difficult without the firm foundation that an overview of the technology provides; leveraging contemporary technology on the modern flight deck, after all, is not unlike what previous generations of pilots have been doing for the past century as new (then) revolutionary (relatively) technology changed the dynamic of airmanship on all previous aircraft models. The description of advanced cockpit technology in this chapter is not simply an overview of wiring diagrams or a tutorial on the terms of art that many may already be familiar with. It is much more than that.

A crucial component of developing a rigorous framework for flying advanced aircraft is an examination of the foundational technology that not only preceded the technology marvels produced by today's airplane manufacturers, but is still an integral part of contemporary flight deck design and engineering. To some this may resemble a history lesson—but beyond that we hope that it portrays the (few) basic problems that advanced technology solves on today's flight deck in a context that keeps every aviator in touch with the ultimate simplicity of the systems dominating aircraft flight deck design.

The directional gyro and artificial horizon that were developed in the late 1920s and early 1930s grew out of technology enabled by the Sperry gyrostabilizer, a product of a company whose origins date back to 1910 in a small factory in Flatbush, New York. The significance of the gyrostabilizer as an ancestor to today's modern flight deck is not simply a history lesson—but instrumental to how crews access and leverage information contained in the FMS as well as in other components of the glass cockpit flight deck.

A brief look at the technology that dominated aircraft cockpits from the late 1920s until the 1980s reveals the elegance of aircraft instrument design, regardless of the level of technology involved. But the singular instrument that allowed the industry to advance beyond the perils of night and instrument flying in the 1920s was the gyro device of the Sperry family, described by T. J. C. Martin in 1929, writing for the *New York Times*:

The gyroscopic stabilizer, as the latest aero-gyroscope is called, has brought automatic control to the threshold of practicability—a threshold, be it said, that bids fair to make it an indispensable unit of equipment in all long distance passenger transport planes. . . . The method by which it masters the plane's controls is an important forward step.[2]

The adaptation of the gyroscopic stabilizer to instrument flight from the rudimentary autopilot (as the gyroscopic stabilizer would eventually be known) and other systems (such as guidance and targeting equipment) would be rapid and effective, essentially providing aviation with a similar revolution in the way pilots and aircrew view and put to use information on today's modern aircraft. But in 1929, the gyroscopic stabilizer did not do much more than allow the pilot to fly "hands off" by turning over the controls to a rudimentary autopilot; there were still challenges that the Sperry engineers had not overcome: nighttime flying, flying in weather, and flying without visual reference to the earth's natural horizon. The initial promise of the Sperry gyroscopic stabilizer was simply that it could be applied to rudimentary autopilots, maintaining level flight and the direction of the aircraft simultaneously. It was an ingenious, yet remarkably simple and elegant device, but it still contained information—*hidden from the crew*—about the horizon and the flight path of the airplane that pilots and crew simply *could not see*.

In *The Triumph of Instrument Flight*, Franklyn E. Daily observes, "Aviation . . . came to understand the essential pilot information was bundled inside the new gyroscopic stabilizer."[3] The gyroscopic stabilizer was, as Daily observes, for practical purposes less than the sum of its parts. Surely it was ingenious—even reliable—technology whose components had to be "unbundled" to propel aviation to yet unforeseen expansion. Heading (direction) information and horizon (attitude) information were both key components of the gyrostabilizer; they simply were not *displayed* in the cockpit for use by the crew. Eventually, three primary instruments would emerge from the gyroscopic stabilizer that together would revolutionize aviation by increasing reliability and dramatically improving safety, all based on the principles of gyroscopes and two of the three Sperry gyrocompasses: the *turn and slip indicator*, the *directional gyro*, and the *gyro horizon* (see Fig. 3-1). All three are still found on even the most modern aircraft, and the genius of the designers and engineers who would make this leap of performance possible was achieved by realizing that the gyroscopic horizon function had to be, as described by Dailey,

. . . *brought out* of the autopilot system in the form of a new display indicator in front of the pilot so that he or she could be in the horizon knowledge loop. This meant that the gyro's sense of the attitude of the aircraft could be given to the pilot, visually. The faith of the pilot in the attitude of his aircraft could be vested in an indicator on his panel. That would correspond to the faith that an actual altitude was coming from the altimeter indicator.[4]

This is perhaps the key insight at the bottom of the modern flight deck: the concept that information must be *brought out* of the automatic systems, as it had been in the 1930s, and presented to the crew in a way that is both useful and reliable, creating a combination of human and machine that is rigorous enough to withstand the forces of nature and predictable enough to formulate a standard approach to operating that can be reproduced on a previously unimaginable scale.

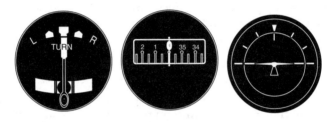

FIGURE 3-1 The cockpit instruments that changed the industry in the 1930s: the turn and slip indicator, left; the directional gyro, center; and the gyro horizon, right.

These three flight deck instruments, together with previously designed pitot-static instruments for measuring airspeed, altitude, and vertical speed, would form the backbone of what would constitute "automated flight" in the decades from roughly 1930 to 1990. Their basic design would see comparatively little improvement during that time, compared to the initial innovation based on Sperry's gyroscopic stabilizer. It would be 60 years before designers and engineers would seize on the breakthrough of digital technology and dramatically reshape the information provided by these instruments, *bringing out* information from these and other aircraft systems that would catalyze an (*almost*) unprecedented revolution in airmanship. What follows on the next few pages is a discussion of the dominant systems at the most basic level, the level at which all operators of aircraft that employ some kind of FMS should understand in order that they might best *bring out of their own FMS* the critical information required for safe, reliable, efficient, coordinated, and *disciplined* flight deck operations.

When Boeing introduced the 757/767 family of transports in the 1980s, it was changing the landscape of transport aviation, fueling competition in an industry that has only accelerated in the time since. Nearly every transport design since their introduction, by all manufacturers, has followed the same fundamental two-pilot, "glass cockpit" layout whose main interface between the crew and flight guidance, navigation, and communication system is the flight management system, or FMS. With the FMS added to the flight deck, the crew had a tool that, when added to the flight instrument family in Fig. 3-1, would allow the crew to see on the flight deck a real-time representation of the aircraft's recent, present, and future flight path across the ground without constant reference to navigation charts and traditional navigation instruments: the magenta line.

At last count, there were over a dozen manufacturers of flight management systems offering the "optimum solution" to the needs of an industry looking to solve many of its problems, primarily through technology. Although the technology solutions proliferate annually at what seems to be an alarming rate (making choices more and more difficult for the buyer), the problems remain relatively unchanged from when they were first identified by research commissioned by NASA immediately after the introduction of the first glass cockpit global transports.

Why New Technology at All

The introduction of a single piece of equipment central to the glass cockpit flight deck, the FMS, is the one factor that has had the greatest impact on flight deck airmanship since the turn-needle-and-ball, directional gyro, and gyro horizon made possible and universalized instrument flight at night and through weather. Pioneering research done by Dr. Earl Wiener and his colleagues in the 1980s and early 1990s identified the driving force behind the new technology back then as consisting of eight primary factors (surely these could have been among the drivers of cockpit technology in the 1930s):[5]

1. Available technology
2. Concern for safety
3. Economy of operations
4. Workload reduction and the crew complement issue
5. More precise flight maneuvers and navigation
6. Display flexibility
7. Economy of cockpit space
8. Special requirements of military missions

These eight primary factors are as much in play over two decades later as they were in 1980s and 1990s (or in 1930!), when organizations were rushing to adopt new technology at a rapid pace. It can be said that these are the very reasons for the introduction of the technology— and should be known by every crew who operates an airplane with an FMS installed. They are the drivers behind the sound business case for the continued proliferation of technologically advanced aircraft worldwide, and they provide an understanding for pilots of the seemingly endless improvements and upgrades to flight deck technology that each will be exposed to over the span of a career in the cockpit.

A Contemporary Model of the FMS, for Pilots (and Mission Crews)

There are many FMSs on the market, and the interface between the pilot and the FMS is different for each; yet just five manufacturers account for roughly 90 percent of the FMSs in use today.[6] And though it is impractical to introduce the salient design differences between every manufacturer's equipment (even the top five), we can offer the basics of the design of the dominant systems, which can be narrowed down to just a few categories employed by most of the major manufacturers. This will allow pilots, crew, and other "agents" in the system to gain a working understanding of the technology as they work to adopt it, not only a working knowledge of the system they operate themselves, but a working knowledge of most of (if not all) the systems operated within their organization or community within the industry. Furthermore, we will show in subsequent chapters that the fundamental requirement of designing, implementing, and adopting *procedures* that guarantee error-free verification of data entry (as just one example of principles shared by all manufacturers) into an FMS is *independent* of the design or the individual manufacturer. We have found in training, operations, and procedures design that understanding both why the technology is present and what constitutes its fundamental logic comprises one of the greatest levers for individuals, organizations, and crews who wish to safely adopt the technology.

Figure 3-2 shows a block flow diagram that represents one of the few basic models of the contemporary flight management system.[7] Daunting at first, and even complex, it demonstrates that among the systems present on the flight deck, one is central to flight deck operations: the *flight management computer* (FMC).

Figure 3-3 extends this concept with a model that we devised primarily to organize and make clearer the role of the automation and the crew. Our model places the *crew* at the top level of the human-machine dynamic and groups the inputs described by Herndon and his colleagues into functional areas that make it easier for aviators to organize, while they build their own knowledge base of the automated flight deck.

An important distinction between Herndon's basic interpretation and our own is centered on the direction of influence of each component in the system and the relationship of the flight deck and/or mission crew to the heart of the automation, the FMS (displays associated with automated major systems such as fuel, electrical,

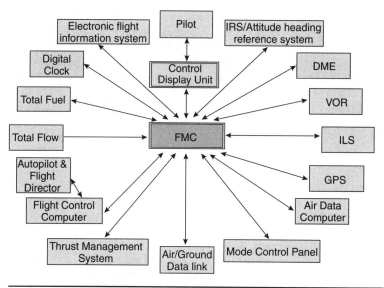

FIGURE 3-2 An earlier model of the basic flight management system.

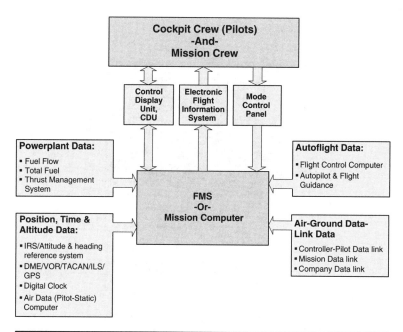

FIGURE 3-3 The fundamental model of the advanced flight deck with relation to the cockpit crew, updated for contemporary airmanship.

pneumatic, and hydraulics are discussed in later in this chapter). Our goal is to represent the FMS in a way that shows the integration of the flight crew more clearly—and depicts the FMS in the way that Klein, Crandall and Hoffman describe it, as ". . . a flight planning organizer, coordinating with other aircraft IT systems to collectively create the auto-flight ability of a technologically advanced aircraft."[8]

Figure 3-4 depicts a "typical" *multifunction control display unit* (MCDU). The MCDU is the dominant interface between the crew and the FMC on most contemporary flight decks; even though there are a variety of manufacturers worldwide, there are comparably minor differences between them (at least in the manner in which crews use them). Although the MCDU is a component of the FMS, because it is the primary flight crew interface, the two terms are often used interchangeably by crews, procedures designers, and training documents. In fact, the MCDU in Fig. 3-4 represents no particular manufacturer's design and has elements of several simply to demonstrate the various functions that the MCDU can control (the trained eye can see that this is a "bogus" MCDU, impractical for use). It can be generally said of all MCDUs that they each will have a display screen to display the myriad "pages" of information that the FMC organizes for the crew's use: line select keys, alphabetic keys,

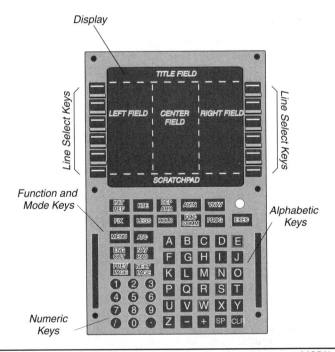

FIGURE 3-4   A generic FMS interface, commonly referred to as an MCDU.

numeric keys and function and mode keys that allow the pilots to select, enter, edit, and otherwise manipulate information essential to managing information on the flight deck.

FMSs have increased in complexity and improved in design over the years since their introduction in the late 1970s. The first-generation FMSs did not provide indications to pilots that they had made entries, nor did they provide a method of verifying that the pilot in fact wanted to make an entry.

Later-generation FMSs added indications to the pilot that entries were *pilot-entered* versus *FMS-calculated.* This was done by using a *large font* on the MCDU display to indicate a pilot entry and using a *small font* to indicate that a value is the result of an FMS calculation. In addition, some FMSs of this generation added an "*insert*" prompt on the respective MCDU "page" that was a method of "asking" the pilot if he or she *did* want to follow through with an entered change in the "flight plan" previously entered in the FMS (a basic, two-step process whose aim is to prevent unwanted or accidental entries to the FMS).

Other FMSs of this generation added an EXEC (execute) button to the MCDU in combination with an "activation" prompt (another version of the two-step process described above). Early-generation FMSs also provided only *two fields* on the display screen for information, a left field and a right field. However, later-generation FMSs provided *three fields* on the MCDU screen for information, a left field, a center field, and a right field (similar to the display in Fig. 3-4). The addition of the center field allowed for greater information display to the pilot as well as greater information input capability. The latest generations of FMS have added color to the MCDU to enhance the presentation to, and the understanding of, the pilots.

As designers and engineers developing FMSs began to take into greater consideration the findings of human factors engineers and researchers, they began to use colors to help flight crews better locate and understand information generated by the FMS. Although there is no universal standard for the use of color even among the most popular glass cockpit designs, the following is a basic overview of the most popular conventions:

- **White** is used for normal data lines and the active route page title.

- **Magenta** is used for active waypoint information and FMS-commanded or "fly to" parameters (waypoint, altitude, airspeed).

- **Cyan** is used for the inactive route page title.

- **Shade** is used for flight plan modifications not yet executed, and for the word MOD in the page title when a modification has been made but not executed.

- **Green** is used to indicate the active state of two- and three-position selectors on a CDU display line.
- On the navigation radio page, **green** is used for VOR frequencies, course entries, and identifiers.
- **Cyan** is used for ADF frequencies.

The different types of FMSs described in this chapter illustrate the variety of ways that designers have found to provide input paths and information fields for the pilots and crew of nearly every type of advanced aircraft in service today. It may seem odd that there is no standard FMS configuration that would apply across the industry; it should, however, be apparent that there are many similarities between the different types of FMSs. After all, there are only so many ways to create the magenta line. For as many types of FMSs as there are, there are at least that many different echelons of pilots who admire a particular type of FMS while disparaging the alternatives. Veteran glass cockpit pilots will likely acknowledge the simple descriptions and limited discussions of the FMS and MCDU in this chapter and perhaps find little value in the technical descriptions we have provided. But for both veterans and novices alike, we consider it essential that pilots and crews consult their aircraft operating manuals for the detailed operating procedures and system features of their respective FMS/MCDU. In fact, most manufacturers are continuously finding new ways for this important component of the flight deck to be used to improve the capability of the FMS to help the flight crew communicate with and manage aircraft information via the FMS through the MCDU. Likewise, we harbor no preference for any type of FMS, only a desire that the basic output and input capabilities be harnessed by the crews to bring out critical flight information, without error, to be used for the safe and efficient conduct of the flight or mission. We think that it is important to know, for purposes of understanding how the FMS, through its various pathways, exercises control over the aircraft guidance and autoflight functions, how they can expect their aircraft to behave when under the influence of the FMS. Chapter 12 discusses FMS and flight control logic in much greater detail as well as how this foundational principle of Automation Airmanship is related to the other eight principles.

Here are a few more general facts that pilots should know about their respective FMSs:

- Gray box FMSs (so named because of the color of the bezel of the unit itself) display a "from" waypoint, while the brown box FMS does not.
- Flight plan changes in the gray box FMS are made on the "Flight Plan" page, while the brown box FMS requires that changes to the flight plan be made on Route or Legs pages.

- The brown box FMS has an EXEC (execute) button, and the gray box does not. This makes it possible for inputs to gray box FMSs to be made directly on the Flight Plan page, or via an insert prompt.

These facts highlight just a few of the sometime subtle differences between manufacturer designs. The arguments for or against any particular FMS design are not as important as the fundamental approach to the technology that every crew should take. Our work in the field with organizations of all sizes, and with missions as wide and varied as theater combat operations, air ambulance, and routine commercial transport, has given us many opportunities to test all nine principles against the most dominant FMS as well as specialty designs. We concluded that no design is resistant to any of the nine principles of Automation Airmanship.

The Working Partners of the FMS

From the basic structure outlined in Fig. 3-3, we know that the FMS is central to the extensive capabilities of advanced aircraft: integrated lateral and vertical navigation, able to be coupled to the speed control system (autothrottles), flight guidance (flight director) and autoflight (autopilot), and drawing precise position information from advanced inertial and GPS systems (whose roots can be found in the Sperry gyrocompass of the 1920s and 1930s). Bringing this information *out of* the FMS for smooth and efficient use by the pilots through the discipline of Automation Airmanship requires additional cockpit components that must be discussed before we leave this chapter of the text.

Like the instruments that formed the backbone of instrument flight in the 1930s, there are, in addition to the FMS, three additional basic components (or "instruments," if that helps you to make the leap) to the twenty-first-century, automated flight deck: the *flight guidance control panel* (FGCP) (or its equivalent, see Table 3-1), the *electronic flight information system* (EFIS) and the display select panel (DSP) (or its equivalent, see Table 3-1). Although these work together and must be mastered by pilots to ensure smooth and easy function, individually they have unique functions, which are tied directly to the FMS.

The *flight guidance control panel* (FGCP) is mounted on the cockpit glareshield and gives the pilots control over the basic modes of flight: airspeed, heading, and altitude. The control knobs function to provide direct input to the instrument displays—most commonly referred to as "bugs" for airspeed, heading, and altitude—and most systems allow for "preselects" to be set as well, through the same knobs. Many FMSs allow for pilots to use the knobs (and some buttons in combination) as a shortcut to inputting data into the FMS. For example, an airspeed can be set on the FGCP via the knob; a button,

Aircraft or Cockpit Manufacturer	FMS: Flight Management System & Input Unit	Flight Guidance Control Panel (FGCP)	Display Select Panel	EFIS: Electronic Flight Instrument System	EICAS: Engine Indicating and Crew Alerting System
Boeing B-777	Control Display Unit (CDU)	Mode Control Panel (MCP)	Display Select Panel (DSP)	Electronic Flight Instrument System (EFIS)	Engine Indicating and Crew Alerting System (EICAS)
Boeing (formerly McDonnell-Douglas) MD-11	Multifunction Control and Display Unit (MCDU)	Flight Control Panel (FCP)	EFIS Control Panel (ECP)	Electronic Instrument System (EIS)	Caution, Advisory and Warning System (CAWS)
Lockheed-Martin C-130J	Communication/ Navigation/ Identification Management System (CNI-MS)	Avionics Management Panel (AMU)	Reference Set/Mode Select Panel	Color Multi-Function Display Unit (CMDU)	Advisory, Caution and Warning System (ACAWS)
Airbus Industrie A-320	Multipurpose Control Display Unit (MCDU)	Flight Control Unit	EFIS Control Panel	Electronic Flight Instrument System (EFIS)	Electronic Centralized Aircraft Monitoring (ECAM) System
Pro Line 21 Cockpit for Business Aircraft	Multifunction Control Display Unit (MCDU) with Cursor Control Device (CCD)	Flight Guidance Panel (FGP)	Glareshield Control Display Panel (GCP)	Integrated Flight Information System	Engine Indicating and Crew Alerting System (EICAS)
Your Aircraft Here					
Future Aircraft Designs Here					

* Not all aircraft and cockpit manufacturers have equivalent systems that perfectly match those of other manufacturers; depending on the age of the technology and the architecture of the system, some cockpits may have systems that combine components which another manufacturer considers separate.

TABLE 3-1 Equivalent Terms for the Primary Glass Cockpit Components*

or "tile" located near the knob can be depressed, and the value set in the displayed window is "sent" to the FMS, bypassing the input steps for accomplishing the same objective by finding the correct input page on the FMS and "typing in" the same value. Every aircraft type and FMS installation differs in some unique way from the others, and interface design is not always identical even across aircraft of the same manufacturer. But it is safe to say that the FGCP (or its equivalent) gives the pilots a ready source to select and input values that provide guidance information to their instruments and the autoflight. Much of the time spent in the early stages of training is devoted to learning and practicing the relationships between the FGCP and the FMS, flight guidance, autoflight, and procedures that are required to be mastered to ensure that pilots can smoothly and effectively bring out the required information to safely fly the aircraft through the entire flight envelope, and in every mode of automation from full manual flight to fully coupled flight.

The FGCP, when coupled with the flight director (the "bars" or "crosshairs" that command the flight path of the aircraft), provides a direct input source for pilots to "drive" the flight path of the aircraft to the desired values of speed, course, and altitude. The FGCP has input paths to select navigation information—source information to the flight director from ground-based navigation aids such as VORs, ILSs, and localizers—and in turn couple this information to the flight director and the autoflight system. It is a real center of action for the pilots. It has been said by a few experienced glass cockpit instructors to pilots making the transition from "steam gauges" to the glass cockpit, "You make your plan in the FMS, but you fly it through the FGCP."[9] Researchers Anjali Joshi, Steven P. Miller, and Mats P. E. Heimdahl describe the flight guidance system as "a component of the overall Flight Control System (FCS). It compares the measured state of an aircraft (position, speed, and altitude) to the desired state and generates pitch and roll guidance commands to minimize the difference between the measured and desired state."[10] A typical flight guidance control panel is shown in Fig. 3-5.

Often located adjacent to the FGCP is another panel that plays an important role in the information presented from the various systems that stream data to the flight deck, the *display select panel* (DSP); see

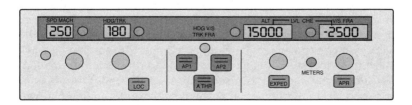

Figure 3-5 A generic flight guidance control panel.

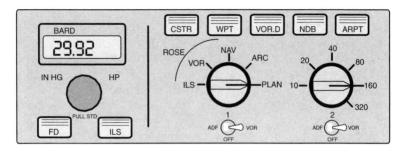

Figure 3-6 A generic display select panel.

Fig. 3-6. On first inspection of the DSP, it appears very similar to the FGCP in that it contains knobs, switches, "tiles" (flush-mounted buttonlike switches), toggles, and display windows. On closer examination in the context of the function of each switch, it becomes obvious that selections of the DSP are different in that they do not impact the flight path of the aircraft in the manner that selections on the FGCP do: the DSP simply enables the display of data—some of it flight-critical—on the situation displays (and the heads-up display, or HUD) and other flight instruments that the crew relies on to maintain situational and mode awareness (this important relationship is the subject of Chap. 9). The DSP is where the crew makes choices about what information is presented for reference, such as waypoints along the route of flight (in addition to those on the flight plan string), fixed navigation aids such as NDBs (Non Directional Beacon) and VORs (VHF Omnidirectional Range), and airports. It is important that individual crewmembers understand the limits of the display—each manufacturer offers many different capabilities unique to aircraft and FMSs—and that this information is *essential systems knowledge* that only enhances an individual pilot's situational and mode awareness.

The DSP is more than mere "window dressing" for cockpit and crew situation displays. The DSP contains controls for setting critical flight data, such as altimeter settings, minimums, and "source" selection for flight guidance navigation aids for different phases of flight (departure, en route, arrival, and approach); and it may contain switches that control the navigation display range, radar and terrain overlay data, and bright/dim controls for each. Although it appears to play a minor role in the ensemble of flight deck components, its expert use is a multiplier of crew performance: the very best crews manage the various settings with ease and always have the panel set to display correct and appropriate information. Set incorrectly, the DSP can seriously limit the ability of pilots to assess critical environmental and navigation information and can lead to a flight path deviation, altitude excursion, or even a clearance violation.

Up to this point, we have not fully "brought out" the information contained in the FMS or generated by pilot inputs to the FGCP to the cockpit crew. Somewhat ironically, however, perhaps because of the name most frequently associated with advanced aircraft—the *glass cockpit*—most readers, whether experienced glass cockpit crews or not, are generally familiar, in concept, with the group of instruments that comprise the *electronic flight instrument system* (EFIS) (or its equivalent, see Table 3-1). The EFIS is simply the collection of electronic displays that replaces the electromechanical instruments that for over a half century formed the backbone of displays in the cockpit. Although it is not the "brains" of the system, or even the backbone of advanced aircraft, the EFIS is the sexy, bright, colorful and dynamic, active center of attention where most of the information is presented, monitored, managed, and confirmed by the flight crew. It truly is the most widely recognized quality of advanced aircraft. But after one masters the "EFIS scan" (just as many pilots mastered the basic instrument scan of less advanced aircraft), its relative importance in understanding the dynamic environment of the technology falls behind the primary focus of Automation Airmanship: knowledge of the logic and command of the system of technology and procedures that brings out and generates the EFIS information for practical use by the flight crew. Commanding this knowledge makes mastering the EFIS a mere exercise in finding and interpreting critical aircraft systems and flight path information.

The main components of the EFIS are the primary flight display (PFD), the multifunction display (MFD), and the Engine Indicating and Crew Alerting System (EICAS). Specialized aircraft may have more functions that can be displayed on these "screens," which vary in number from just one to as many as eight or more, that can be configured according to the desires of the crew. For variations of these terms across a select handful of manufacturers, see Table 3-1 and apply the substitute terms to your own aircraft, if they apply, while you continue with this chapter.

The PFD, as its name implies, is central to the pilot's scan, directly in front of the pilot's forward view, and contains virtually every real-time piece of flight data that the pilot needs to *fly the aircraft now*—from basic information on aircraft attitude, airspeed, heading, and altitude to FMS and FGCP generated information about other critical data related to the configuration and modes in which the automation and autoflight systems may be operating. Depending on the particular individual design, the PFD can also contain "trend" information (e.g., where the aircraft speed will be in 10 s given current flight path dynamics), real-time wind information (direction and speed), the status of autoflight and flight guidance modes (through areas of the PFD referred to commonly as the *flight mode annunciators*, or FMAs) and even weather, wind shear, ground proximity, TCAS, and other information generated by a variety of sensor and display systems. It is—if "avalanche of information" is a metaphor for the glass cockpit—the place where the leading edge of the cascading flow meets the fall line, and either passes harmlessly over or completely consumes what lies in its path (depending on how well the crew has managed the information input, the physical flying of the airplane, and their awareness of the relative importance of all the data in view).

Normally not integrated into the EFIS are the standby instruments—instruments that generally receive their source information from the aircraft's pitot-static system and independently powered attitude source and directional gyro. We mention them not because they can sometimes appear as or be grouped with the primary flight instruments on the EFIS. We mention them because knowing how they are powered and in turn generate basic flight information is foundational in understanding cockpit crew actions when the EFIS, or components that normally generate reliable flight information displayed there, fail. For whatever aircraft you fly, you should know the details of the standby instrument system and when possible and practical, practice using them.

Another component of the glass cockpit that an increasing number of contemporary aircraft are being equipped with is the *heads-up display,* or HUD. A complementary technology to the HUD is the *enhanced vision system* or EVS (or some equivalent term unique to the specific manufacturer) that enables the crew to "see" through conditions of low visibility for increased situational awareness. Installations and capabilities of both these systems can vary widely, but the fundamental display information is pulled from the same sources that generate critical flight data displayed on the PFD. This simply allows the pilot to "overlay" critical flight data on the real world as seen through the cockpit window without scanning the forward instrument panel. Advantages include enhanced operations in low visibility and assistance in takeoff and landing operations during night and instrument conditions, depending on the installation and certification requirements of the individual organization, airline, or flight department. The best place to begin to understand this cockpit equipment is in the manufacturer's guidance or the operational guidance provided by the respective operator. These and other related systems are finding their way into cockpits at a rapid rate; but the principles we outline in this book are easily adapted to operations with HUD and EVS, since they represent just another way in which critical flight information is brought out of the various aircraft sensors and systems for use by the crew.

No matter what technology pilots in different aircraft and cockpits use to generate a visual representation of the aircraft flight path or the magenta line, it must be accomplished through a discipline whose goal is to eliminate input errors, followed by an equally important set of actions whose goal is to accurately verify that the information is correct. In Chap. 4 we discuss the first of the nine principles of

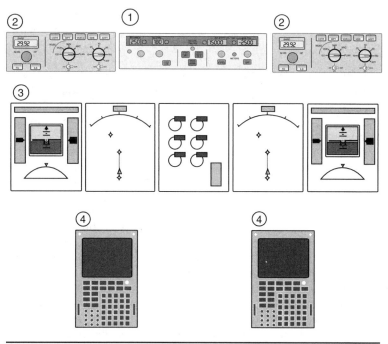

Figure 3-7 The four components of the twenty-first-century automated cockpit: (1) the flight guidance control panel (FGCP); (2) the display select panel (DSP); (3) the electronic flight instrument system (EFIS), and (4) the flight management system (the primary interface), the FMS or MCDU.

Automation Airmanship that allow the most inexperienced glass cockpit crews to bring out the essential information contained in the integrated FMS. Like the earlier diagram we presented in Fig. 3-1, which formed the backbone of the instrument flight deck that revolutionized aviation in the early twentieth century, the above components of the modern flight deck combine to form the foundation of fully integrated, automated flight in the twenty-first century (see Fig. 3-7).

Automation Alone Is Not the Answer

In 2005 we first stepped in front of a large group of pilots, aircrews, researchers, safety experts, and other members of the aviation industry and suggested that pilots of glass cockpit aircraft of all types be trained differently than they have been over the previous 60 years. We built a case then for emphasizing in training and evaluation—from *ab initio* pilot training through currency and continuation training for all pilots over their entire careers—a serious grounding in the logic of the automated systems of the aircraft that they are qualified in. In fact, some organizations have blended this knowledge into training curricula as experience has grown through the past decade.

Yet, across the industry, there is no consistent schema for introducing, instructing, and evaluating the logic of automation. We illustrate this nearly every time we present our material in front of our colleagues across the industry with a concise summation of technology from a widely known and respected designer of technology applications, Microsoft founder and chairman Bill Gates. According to Gates, "The first rule of any technology used in a business is that automation applied to an efficient operation will magnify the efficiency. The second is that automation applied to an inefficient operation will magnify the inefficiency."[11] Audiences all over the world universally share the response to this observation by Gates.

Nobody is ever surprised by Gates' first assertion about applying technology to an already efficient operation; but when we recite the second assertion, of this by now notorious quip, that *an inefficient operation adopting technology is verily a recipe for disaster*, then heads in the audience really begin to nod in agreement, and pencils raise to record the citation for future reference.

It seems that everyone has an example of this principle from personal experience. It is no surprise, then, that we are often asked to provide our experience in promoting efficiency across the organization by organizations in the midst of adopting advanced aircraft. At the organizational level, a variety of institutional structures must exist to enable efficiency, allowing organizations to realize the promise of new technology (we deal with this directly in Chaps. 13 and 14). But at the point of actual service of the new technology—in the aircraft itself— the first step toward creating an efficient operation on the modern flight deck is to develop a firm understanding of the technology architecture itself, embodied in the flight management system (FMS), known in some aircraft as the mission computer (MC).

Many books have been written about the "madness" of machines and automation gone awry, creating maelstroms of failure which have in some cases led to massive loss of human life and property. This book is not one of them. Yet we need to be certain that the reader understands our fundamental approach to the frontier of performance where the human operator interacts with the machine; we do not propose as a solution to an increasingly complex world, in which Information Technology (IT) is intended to help us organize and manage complexity, a philosophy of "all automation, all the time" or even "all automation, most of the time."

As a foundation for those whose professional (or even personal) lives are dominated by complicated systems, we propose that they be informed by a fundamental understanding of the logic of the automated systems and their interactions with other systems. In short, every pilot should seek an understanding at first contact with her or his new airplane of the heart and soul of the technology—the logic that drives the overall system itself. Ideally, this is provided by the training system that supports every pilot's transition to a new aircraft. If a training system lacks this vital component of modern airmanship, individual pilots should pursue it themselves, for this is indispensable, front-line information: the knowledge of the logic that comprises the control and management of the flight path and automated systems which give modern aircraft cockpits their distinctive elegant design and smooth function. In this chapter we have provided a basic primer for understanding the most common FMSs, the basic components of the glass cockpit flight deck, and the fundamental relationship between the technology and the crew—information we believe is lacking in many publications that deal with automation. With this knowledge and the information in your aircraft flight documents, you will be able to absorb the next few sections of this book and begin to experience better, safer, and more effective outcomes on every flight.

Notes and References

1. Franklyn E. Dailey, Jr., *The Triumph of Instrument Flight,* Dailey International Publishers, Wilbraham, Mass., 2004, p. 57.
2. Franklyn E. Dailey, Jr., *The Triumph of Instrument Flight,* Dailey International Publishers, Wilbraham, Mass., 2004, p. 57.
3. Ibid., p. 57.
4. Ibid., p. 59.
5. Earl L. Wiener, Barbara G. Kanki, and Robert L. Helmreich, *CRM in High Tech Cockpits, in Cockpit Resource Management,* Academic Press, New York, NY. 1993, p. 206.
6. Albert A. Herndon, Ralf H. Mayer, Randal C. Ottobre, and Gregory F. Tennille, "Analysis of Advanced Flight Management Systems (FMSs)," MITRE Corporation: Center for Advanced Aviation System Development. McLean, VA. July 2006.
7. Ibid.
8. Beth Crandall, Gary Klein, and Robert R. Hoffman, *Working Minds.* MIT Press, Cambridge, Mass., 2006, p. 260.
9. Bill Hutzell, a retired Captain for Braniff and Piedmont airlines and a long-time glass cockpit flight instructor, has been providing thousands of aspiring glass cockpit crewmembers with crucial insight like this for over two decades. It is from Bill that the authors first heard this simple and insightful advice.
10. Anjali Joshi, Steven P. Miller, and Mats P. E. Heimdahl, "Mode Confusion Analysis of a Flight Guidance System Using Formal Methods," University of Minnesota, Rockwell Collins and the NASA Aviation Safety Program and the Langley Research Center under Contract NCC-01001, 2003. p. 3.
11. Bill Gates (1955–), entrepreneur and founder of Microsoft.

PART II

Organizing the Advanced Flight Deck for Optimum Crew Performance

The FMS is a flight-planning organizer, coordinating with other aircraft IT systems to collectively create the auto-flight ability of a technologically advanced aircraft.

—Crandall, Klein, and Hoffman, 2006[1]

CHAPTER 4

The First Principle: Planning

If we didn't have goals, we wouldn't do anything.
—Dietrich Dörner, 1996[2]

Experience Alone Is Not Enough

Experience does not always equate to expertise, and it does not guarantee expert performance. On November 22, 2004, two pilots flying a Gulfstream III, call sign 85VT, from Dallas to Houston, Texas, faced what under any circumstances would be considered a routine mission: to make a short flight of less than 250 mi between large metropolitan airports to embark passengers (including the former President, George H. W. Bush) and then to a destination just several hours away, which on this day was in Ecuador. The combined experience of the crew—the Captain (age 67) with almost 19,000 hours of total flight time and 1000 hours in G-II/III aircraft and the First Officer (age 62) with over 19,000 hours of total flight time and 1700 hours in type—equated to more than two full careers of flying experience to rely on. The weather in Houston was near the minimums for their operation, with low ceilings and visibility reported, yet well within the capabilities of the airplane, and certainly within the ability of one of the most "experienced" crews of any accident of its type ever to be investigated by the National Transportation Safety Board. In the short flight which lasted from 0530 to almost 0615 local time, this highly experienced crew would take a routine trip in a highly automated aircraft and experience the worst possible outcome, due in part to what the NTSB would later report as ". . . the flight crew's failure to select the instrument landing system frequency in a timely manner and to adhere to approved company approach procedures, including the stabilized approach criteria."[3] In separate but similar circumstances, less experienced crews operating equally capable aircraft flew to Houston on that same day, within the same time frame, and their flights were soon forgotten as having happened without incident. How can it be that this crew was not successful? In

49

the absence of detailed accounts of the preparations they undertook for their flight on that day, we can only speculate on the preparations that they undertook—or failed to complete—before their departure to Houston. The conditions present in this accident are so common, and contain so many aspects of the demands of modern flight operations, that the accident serves as an ideal example of "a good day gone bad," whose lessons can be a beacon of action for the many thousands of crews who will face similar circumstances many times over their own careers. The planning considerations of this flight are merely the beginning of how events would later go tragically awry, and they serve as an example of how a "team of experts" does not always result in "an expert team."

The narrative of almost every accident, incident, or disaster always includes a discussion of the activities leading up to the mishap, in a search for evidence of where individuals or systems acted or failed to act, thereby contributing to the event. As we begin the narrative of the nine principles of Automation Airmanship, we choose to start before the airplane leaves the blocks—and even earlier, prior to the crew's arrival at the aircraft. In this discussion, our focus is on the planning that expert crews undertake in flying any type, size, or brand of advanced aircraft. We are *not* talking about what has become known across the industry as *crew resource management* (CRM) activities related to preparing for flight duties—the likes of which include proper rest, exercise, and personal habits that limit the intrusion of distractions during hours of flight duties requiring peak performance. Rather, our emphasis is on the approach that the best crews take leading up to starting the engines that allows them to optimize the performance of the powerful combination of well-trained pilots with modern flight deck technology over the course of their flight, mission, or duty day. There is an approach that experts take that is indifferent to whether the problems they encounter are routine or involve significant and unusual circumstances. If you choose to make these practices part of your routine preparation habits, it would be but one way to organize your increasing expertise.

We have stated our belief that many professional aviators know a great crew or crewmember "when they see one . . ." but find it difficult to quantify the qualities that separate the truly elite from the broader population of pilots and aircrew. Just the same, the legacy of modern aviation is replete with stories and reports of truly outstanding feats of airmanship; many of us are familiar with US Air 1549 (New York, 2009) and United 232 (Sioux City, 1989), among dozens of others that have not been judged by contemporary media to contain sufficient drama or public appeal to be widely known outside of aviation. In examining the narratives of many of these accidents, the image of heroic pilots begins to blend with the vital roles played by their crews and other agents outside the cockpit, on the ground, or in other aircraft. Undoubtedly, despite the self-effacing testimony of the central figures of these high-profile events to the public and investigators alike that "I was just doing what I was trained to do" and "I couldn't have done it without the help of my crew," there is much evidence of a pattern of preparation and professionalism that many accidents with poor outcomes lack. Our observations and the research both bear this out.

It is universally accepted by pilots and crews across all of aviation, and increasingly across many other high-risk industries, that *situational awareness* (SA), defined as "the perception of the elements in the environment within a volume of time and space, the comprehension of their meaning, and the projection of their status in the near future,"[4] is a component of almost every successful flight activity, routine and otherwise. Indeed, it is noted as a "Capstone Outcome" in the Historical Airmanship Model® developed by our colleague, Dr. Tony Kern. Much of the time spent in both training and line operations is devoted to building and maintaining SA, and as one would expect, research shows that there is a difference between the SA of a novice and that of an expert. We delve into SA later in this book, as a component of a separate principle of Automation Airmanship, but we need to address the importance of planning in developing SA, and the difference between experts and novices as they build SA at the earliest stages of the flight, in planning for the flight itself.

How Experts Plan

We are confident in our assessment of how to plan as an expert does when operating complex aircraft, but in this case we are assisted by a solid body of research that demonstrates how important it is to adopt the skills of the best performers to improve our own. As the adoption of advanced aircraft was well established and accelerating in the late 1990s, veteran researchers Carolyn Prince and Eduardo Salas discovered two critical differences between a study group of pilots that included less-experienced, general aviation (GA) pilots with an average experience level of 720 hours, more-experienced airline pilots with an average level of experience of 6036 hours, and check airmen, considered to be among the best commercial pilots with an average experience level of 12,370 hours.[5] First, as experience increased, so did preflight preparation: the best pilots spent more time and obtained better information through planning specific to the upcoming flight. They gathered as much information as possible during the time devoted to this phase of flight, including details specific to flight conditions, conditions of the aircraft, and made provisions for contingencies that the less-experienced pilots did not. Second, the check airmen were more likely than the other two groups to focus on what researchers classify as "level 3 SA" (a mental model of future states based on a projection of current conditions and its dynamics),[6] by developing a pattern of proactive planning, organizing and understanding a larger amount of information, and understanding the relationships among various factors affecting the outcome of the flight or mission. By doing so, these pilots pulled away from their less-expert colleagues (GA pilots and average airline pilots), increasing the safety margin of their crew and the organizations they fly for. Clearly, the lever that level 3 SA provides should be reason enough to engage in planning that seeks better, more-detailed information that can be used to reliably shape favorable outcomes. In the United States in 2009, ten years after the findings of this research were published, GA pilots accounted for 474 fatalities compared to 50 fatalities among scheduled airlines, and 7.2 accidents per 100,000 flight hours—nearly *50 times* the accident rate (0.149 per 100,000 flight hours) of the more experienced crews of scheduled airlines.[7]

To demonstrate this visually, without sifting through pages of research and volumes of data, we can again turn to chess enthusiasts and then draw a similar model that works within aviation, in much the same way. In the example below, you only need to know the basic moves of chess to immediately see how fast and efficiently the expert could dispense with a novice in facing identical conditions. You can draw the parallels to aviation as easily as we can, but we will provide an example of expert preflight planning before showing you deeper insights related to this vital skill in operating complex, advanced aircraft (large or small, whether you are a GA pilot flying the latest in light aircraft, or a veteran of decades of glass cockpit flying for a corporate flight department or major airline). In this example, K. Anders Ericsson explains the distance that separates a club from an expert chess player.[8] Center your attention on the white pawn on c4, and assess your options. Then compare them to the potential moves of the prototypical expert. It is your move. See Fig. 4-1.

You can see that the expert, on the right, may begin planning the next move in the same way as the novice on the left, given the same scenario. But the expert performer, knowing the importance of planning for contingencies and through experience, "sees" more options than the novice.[9] The expert is constantly prepared to think past the first, second, and even third steps, as she or he assesses the dynamic nature of the likely moves which will force alternatives, bring him or her and the entire crew toward level 3 SA. Here is an example of an expert approach to flight and mission planning, as it might be represented in a map as in Fig. 4-2.

You do not have to be a chess master if you want to develop level 3 SA, but it helps if you can think as one. Our assessment of what the best glass cockpit pilots are engaged in doing at the earliest stage of each flight or mission, and why, is summarized in the description of Automation Airmanship principle 1:

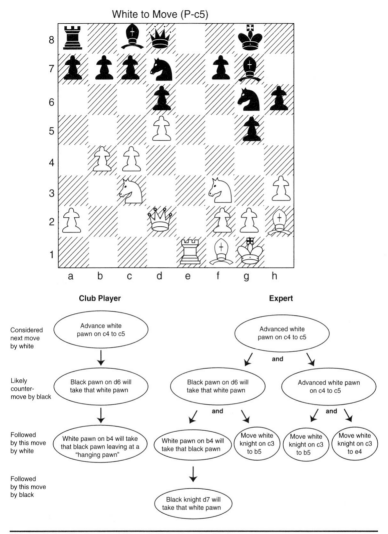

FIGURE 4-1 Novice and Expert Thinking in Chess.

Because we want to capture the actions of the best crews and build a formula for success around them for others to implement in their own operations, we have to understand how expert crews approach flight planning. Both research and anecdotal observation point toward expert performance as having a component of advanced planning in which the experts involved seek out information and create alternatives that novices do not. We are reminded that the technology that exists in the aircraft is a tool for planning and organization, and as such it is included in the processes that experts

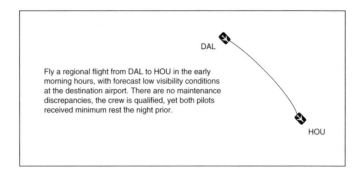

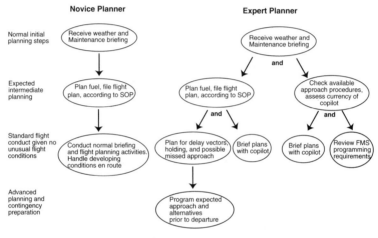

FIGURE 4-2   Novice and Expert Thinking in Flight Planning.

devise when formulating their plans for successful outcomes. Often, the necessary first step is to use the first stages of the crew briefing to assess the proficiency, readiness, and competency for the flight or mission. An underevaluation or inaccurate assessment of these three factors will, at a minimum, increase the workload of the more prepared crewmember(s), making task management more difficult for every member of the crew. We have learned from our interaction with both experienced and inexperienced crews, civil and military, that when it is well known in advance how well a crewmember can be expected to interact with the automated flight systems to achieve a desired outcome (how well he or she can handle, for example, the necessary steps to program an RNAV (Area Navigation) arrival and subsequent ILS approach), then each crewmember can adjust to manage and meet those expectations. A high level of automation on the flight deck makes preflight and planning different from the same phase of flight in less complex, legacy airplanes. The best pilots know the level of competence that they can expect from their fellow crewmembers with respect to their ability to manipulate the automated systems on the flight deck. And the best pilots know this before they actually arrive at the aircraft, in almost every case. They also know that once the opportunity passes to conduct this important step of flight planning, the fast-paced and dynamic environment of real-time flying can swallow them whole.

Detailed Planning in Automated Aircraft

As one airline pilot who had undergone the transition from a less complex aircraft to a similar version with increased flight deck technology once quipped, "DC-9 pilots are men of steel on ships of wood. We set our sails and go. In the MD-88 you're always fiddling with something."[10] Using flight planning as a lever includes understanding planning itself. Under the heading of this chapter we are reminded of the purpose of having goals in providing the impetus for action. When setting goals, we are always doing one of two things: defining an outcome we want to happen or identifying a condition we want to avoid. We want, for example, to arrive at our destination on time. We also want to avoid areas of severe weather, conditions and periods during the flight when workloads are inappropriate and unmanageable, and literally hundreds of other negative outcomes that could possibly beset the flight. Deitrich Dörner describes planning in his book *The Logic of Failure: Recognizing and Avoiding Error in Complex Situations* in this way:

> What is planning? In planning we don't *do* anything; we just consider what we *might* do. The essence of planning is to think through the consequences of certain actions and see whether those actions will bring us closer to our desired goal. If individual actions will not achieve our purpose, we have to lay out sequences of actions.[11]

Goals are specific or general, and in planning we often decide to further define a goal when we have access to information that permits it; or we decide to keep a goal broad and open-ended, allowing us to gather more information. The best pilots and crews understand this relationship and take it forward from preflight into the mission itself. Chapters 5 and 6 explain how this happens through other Automation Airmanship principles.

"Laying out sequences of actions" is a heavily loaded phrase in the context of what we are doing as glass cockpit pilots, setting our flight goals and planning the steps along the way. In less automated, legacy aircraft without FMS flight guidance, vertical navigation capability, and large navigation and situation displays, goals can be left more general and planning can be less detailed (pilots and crews can, in fact, "set sail and go"). After all, it makes no sense to brief and plan every contingency at the arrival airport—there is no way one can hold that much useful information without seriously handicapping other cognitive functions. Additionally, there is no way conventional cockpits can be set up in advance for the en route segment and the arrival and approach procedure in the way that the fully integrated, automated flight deck provides. The commitment to a new approach to setting goals and planning usually develops in training, when the pilot realizes the information that the advanced systems can hold—well in advance of the arrival in the terminal area of the destination—without having to remember every detail. The convenience of setting up multiple contingencies through the FMS is the most compelling argument to increase the time and effort spent planning. Yet even the very best pilots understand the problem of "scale" in planning, a trap that has been laid for decades at the feet of planners in every domain: "We can make plans that are too crude and plans that are too detailed. The trick is to plan with an appropriate degree of detail."[12]

As every aviator knows, planning is not the magic bullet for avoiding in-flight occurrences that can alter an hour's worth of excellent planning. Yet in assessing the crew, the environment, and the equipment, crews can be better prepared for whatever contingency befalls them. The other component of flight and mission planning is the assessment of crew duties with respect to expected responsibilities—the setup duties required of a precision approach in low-visibility conditions, for example—and how those duties will be carried out in-flight is easily accomplished in planning. Assessment of equipment has been done by crews for years; crews of automated aircraft rely on sensor systems much more heavily than they did when operating less complex aircraft; crews that notice, discuss, and include in their planning the limits to the operation placed on them by inoperable systems will formulate better plans, at the appropriate level of detail, and have better outcomes as a result.

Checklist for Success: Planning

- ☑ Conduct a preflight/premission assessment of crewmember proficiency, readiness, and competence in operating automation.

- ☑ Assess aircraft equipment status, considering the impact of degraded or unavailable systems on the planned flight.

- ☑ Brief overall expectations for automation use during high-workload phases of flight, including any exceptions to SOP and intentions, if any, to use less automation for manual-flying proficiency.

- ☑ Brief augmented crewmembers, if any, on their expected backup and monitoring duties.

Notes and References

1. Beth Crandall, Gary Klein, Robert R. Hoffman: *Working Minds: A Practitioner's Guide to Cognitive Task Analysis*. The MIT Press. Cambridge, MA. 2006. p. 160.

2. Dietrich Dörner, *The Logic of Failure: Recognizing and Avoiding Error in Complex Situations*, Metropolitan Books, New York, 1996, p. 49.

3. National Transportation Safety Board. Crash During Approach to Landing, Business Jet Services, Ltd. Gulfstream G-1159A (G-III), N85VT. Houston Texas. Aviation Accident Brief/AAB-06/06. 2006. Washington, DC. p. 21.

4. Mica Endsley, Expertise and Situation Awareness. In *The Cambridge Handbook of Expertise and Expert Performance*. K. Anders Ericson (ed.), Cambridge University Press, Cambridge, England, 2006, p. 634.

5. Carolyn Prince and Eduardo Salas. "Situation Assessment for routine flight and decision making." International Journal of Cognitive Ergonomics, 1(4), pp. 315–324 .

6. Mica Endsley, Expertise and Situation Awareness. In *The Cambridge Handbook of Expertise and Expert Performance*. K. Anders Ericson (ed.), Cambridge University Press, Cambridge, England, 2006, p. 634.

7. National Transportation Safety Board, Review of U.S. Civil Aviation Accidents 2007–2009: NTSB/ARA-11/01PB2011-113050, Notation 8290. Washington, DC. Adopted March 31, 2011. pp. 1–3.

8. K. Anders Ericsson, "Protocol Analysis and Expert Thought: Concurrent Verbalizations of Thinking during Experts' Performance on Representative Tasks," *The Cambridge Handbook of Expertise and Expert Performance*. K. Anders Ericson (ed.), Cambridge University Press, Cambridge, England, 2006, p. 234.

9. Ibid.

10. Earl Wiener, *Crew Coordination and Training in the Advanced-Technology Cockpit*, Academic Press, New York, NY. 1993, p. 199.

11. D. Dörner, *The Logic of Failure: Recognizing and Avoiding Error in Complex Situations*, Metropolitan Books, New York, 1996, pp 153–154.

12. Ibid., pp. 49–56.

CHAPTER 5

The Second Principle: Briefing and Debriefing

If you are sweating too much before the flight, you haven't asked enough questions. If you are not sweating just a little during the flight, you are not attentive enough. And, if you are not sweating out your remaining questions with all the experts that you can think of after the flight, you may never find that very beautiful pearl in the pig litter.
—Corwin H. Meyer, Grumman test pilot, 1942–1978[1]

For over 20 years, the concept of providing a comprehensive briefing and debriefing has been a cornerstone of crew resource management (CRM) programs across the industry. It should be no surprise, then, to find that this cornerstone skill is magnified in importance aboard aircraft whose complexity far outstrips that of their predecessors, with capabilities only dreamed of just a decade ago and operated by smaller (and often less experienced) crews. In upcoming chapters we will discuss the direct methods that expert crews use to leverage the FMS and other IT systems on board, skills that would be less accessible— or even unavailable—without the "setup" of a good briefing. Similarly, in a cockpit with a potentially overwhelming volume of information both entering and leaving (often without the knowledge of the crew) over the duration of any flight, and the tireless calculus of the flight management system (FMS), with its cool, emotionless synthetic memory, failure to have a comprehensive recap of a flight just ended would be an irresponsible waste of valuable experience. We do not offer a review of how to brief and debrief a typical flight in just any aircraft; rather we demonstrate how to improve and adapt these skills to the context of the advanced cockpit. Knowing how expert glass cockpit crews use their briefings to enhance the performance in the demanding environment of technology-centered flight deck operations is a front-line skill that every glass cockpit crewmember can master. We have put briefing and debriefing together for the simple purpose of connecting these two skills to each other—and keeping them tightly coupled, so that crews will be able to share insights and opportunities to learn with the same rigor with which they conduct clear and concise briefings that are compatible with the automation.

In Chap. 4 we considered the plight of the crew of Eight-Five Victor-Tango as they fought to keep their airplane on the desired flight path during a routine approach and landing in reduced visibility. We used the opportunity to build the case for planning—planning that is detailed enough to outline the requirements and assign responsibilities in achieving goals, but not so heavy that it weighs the crew down—an activity which is closely related to briefing and debriefing, but is limited almost exclusively to preflight. In mission planning we talk about what we hope to do; during briefings we review the detailed steps of carrying out the plan, modifying the plan or even reformulating the plan. The big difference is that in the latter setting we have at our fingertips, and before our eyes, the tireless and dutiful components of the aircraft's automated systems, the flight management system and its accompanying displays and interfaces: briefings literally help bring our plans to life on the automated flight deck. Because of their role over the entire operational phase of flight, briefings often span the entire flight or mission; and frequently briefings overlap with debriefs and recaps, allowing no detail to escape its place in executing the plan, or in allowing any detail to escape capture when an occurrence (good or bad) can be leveraged into enhancing future performance.

The placement of a complete briefing, compatible with the approach being flown and the technology on the flight deck which allowed for a full display of the lateral navigation track, would have certainly served to intervene during the events that would unfold as Eight-Five Victor-Tango approached Houston Hobby airport on the morning of November 22, 2004. From CVR records, we know that the first officer conducted the majority of the briefing, but that company policy required the captain (the pilot flying the approach) to conduct the briefing. In addition to this clear deviation from company SOP, the first officer's briefing lacked critical information regarding the final approach fix (FAF) altitude and the required aircraft configuration. Following the briefing, as the aircraft was approaching its destination, the captain directed the first officer to enter waypoints for the approach in the FMS: CARCO, an intermediate fix 14.3 nautical miles from the runway; ELREN, a stepdown fix for the localizer approach, 7.3 nautical miles from the runway; and EISEN, the FAF, 4.3 nautical miles from the runway. See Fig. 5-1.

The Hobby VOR (on the airport) was previously entered into the FMS, but likely removed by the first officer after a discussion with the captain about whether or not it was required for the approach (it was not). This would have left EISEN, the final approach fix, as the last waypoint in the FMS string entered by the crew, and the last waypoint displayed on the multifunction display (MFD, the display in the cockpit where waypoints in the FMS are listed in order).

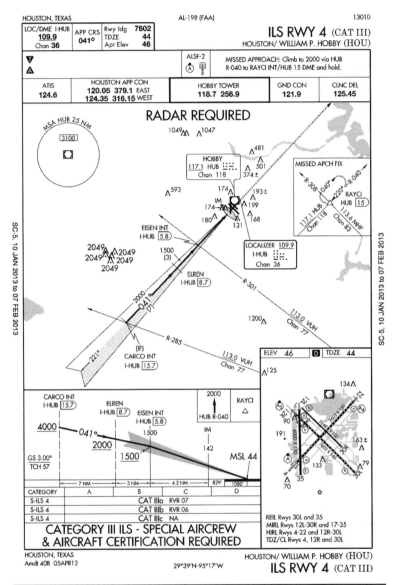

FIGURE 5-1   The ILS to runway 04 at Houston Hobby airport, Texas.

The report concluded that it is possible that this sequence of actions—disconnected from the first officer's previous briefing—may have induced a fatal confusion for the crew:

The MFD only displays a chronological number for each approach waypoint; therefore, it is possible that the flight crew forgot that the first officer removed the [VOR] waypoint from the FMS, causing them to mistakenly believe that the last waypoint displayed on the MFD (EISEN) was the airport.[2]

All experienced pilots of advanced aircraft know that on an ILS approach, the FMS and its accompanying depiction of the aircraft lateral and vertical flight path are a secondary reference for the approach, and that the primary flight guidance is the ILS equipment and its display components. If the data are properly entered, they should be identical. This fact would certainly have been discussed in a proper approach briefing, making the first officer's focus on the FMS stand in stark contrast to the original plans to *fly the ILS*. See Fig. 5-2.

What constitutes the skill of briefing and debriefing that the best-performing crews invoke on every flight is summarized as follows:

The Vital Role of the Briefing

In the 1990s Gary Klein and his associates sought to explain the most subtle aspects of executive decision making in high-stakes endeavors,

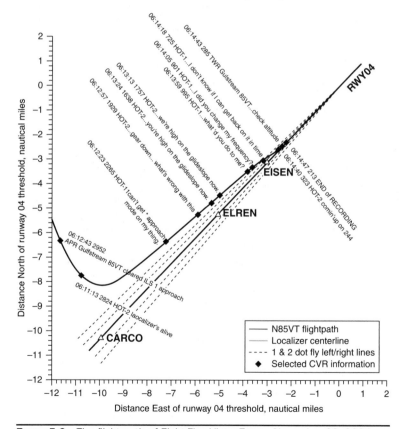

FIGURE 5-2 The flight path of Eight-Five Victor-Tango, November 22, 2004.

a task that had eluded researchers for decades and, to some extent, still does. Yet what he and his associates learned by studying groups of military commanders, firefighting commanders and nurses in neonatal intensive care units, has immediate application for crews of advanced aircraft. The research Klein did in this field grew into what is now known as *naturalistic decision making* (NDM). In his seminal book on the subject, *Sources of Power*, he articulates a variety of "powers" that expert performers routinely rely on in the face of demanding, often life-threatening conditions. (His book should be on the reading list of any professional who interacts with others in demanding, dynamic, time-critical situations.) One of these is what Klein calls the "power to read minds." We want to transfer some of these concepts directly into all aviators' battery of front-line skills and inoculate them against confusion, indecision, and ambiguity. Think of briefings in this way: "We've made a plan, and now this is how we carry it out." Or, according to Klein,

> Whenever we make a request—ask for an errand or give a command—we need the person to read our mind. To make this possible, both parties have to extend themselves. The person making the request can help by specifying the intent behind the requests. The person trying to carry out the request has to imagine what the other person really wants, to handle all the details that did not get explained.[3]

The confusion which the first officer brought to the cockpit of Eight-Five Victor-Tango (permitted by the captain's lack of a comprehensive briefing) confounded the possibility that either pilot would be able to "read the mind" of the other. In fact, the relationship was, for at least a few minutes, reversed, with the captain attempting, through his indirect comments about the approach programming by the first officer, to piece together what his colleague was trying to accomplish. Meanwhile, the primary navigation aid (the ILS) was not only operational at the field, but also unavailable until after the aircraft had commenced the approach and it was finally tuned by the first officer. The actions of this crew make clear that in a rapidly changing, dynamic environment, there exists the problem of "guessing intent" which Klein seeks to minimize (or eliminate, if possible) by explaining the purpose of establishing "intent" in the first place, and then offering a formula that high-performing crews can (and some already do) use to establish intent within the crew.

Functions of Communicating Intent

1. Promote independence.
2. Improve team performance by reducing the need for clarifications.

3. Detect deviations from the assumptions made by the leader.

4. Catch errors in advance and anticipate problems.

5. Promote improvisation.

6. React to local conditions without having to wait for permission.

7. Recognize opportunities that were not part of the plan.

8. Set priorities in order to make trade-off decisions.

9. Continue beyond the outcome without having to wait for the next order.[4]

No captain, aircraft commander, or mission commander envisions a scenario where the crew is expected to act independently of others (in the way the crew of Eight-Five Victor-Tango did for at least a few critical moments), but they do expect crews to be able to think and act in the context of the well-briefed plan. For crews to apply this list and build briefings that communicate intent, it must be converted to action steps that actually deliver the desired outcome. These action steps then need to be reliably reproduced for countless situations, and crews need to be able to apply them without contemplation. Klein offers seven types of information that leaders can present to help the people receiving the request to understand what to do.[5]

1. The purpose of the task (the higher-level goals)

2. The objective of the task (an image of the desired outcome)

3. The sequence of steps in the plan

4. The rationale for the plan

5. The key decisions that may have to be made

6. Anti-goals (unwanted outcomes)

7. Constraints and other considerations

To formulate a structure for issuing instructions (or conducting briefings) derived from his research into the performance of experts in high-stakes situations requires a fundamental leap for all glass cockpit crewmembers—once made, every briefing that follows over a lifetime of flying will permanently improve performance—and it only requires a small amount of practice to master. And it is far easier for crews of advanced aircraft than for professionals in other high-stakes industries.

Aviators of even limited experience can read the list above and see its applicability to flying, and so we have modified Klein's list of seven into a more concise formula specifically tailored to the highly automated flight deck. In the early stages of training, most pilots are taught the fundamentals of briefing departures and approaches, and later they extend the same framework to taxi routes, climb, en route and descent segments, and other essential tasks. During this time, 90 percent or more of the briefing content comes directly off the charts used for navigation, with some reference to the instruments and their settings. Unfortunately, transferring this same methodology to the glass cockpit flight deck can be hazardous, or even fatal. Before the briefing of a procedure in the automation environment, the question to be answered is not "What's the procedure look like on our charts?" but rather "What's the airplane going to do as a result of our programming the procedure?" In fact, Klein's research points to a dramatic yet elegant fact that has a universal application to all automated aircraft: "The concept of Intent can be applied to equipment as well as to people."[6]

Once pilots and crews of automated aircraft grasp this notion, they are prepared to dramatically improve their performance and leverage technology in the way that the designers and engineers initially imagined. What this translates to is that in glass cockpit aircraft with worldwide FMS database capability and automatic tuning and identification of navigation aids, pilots no longer have to manually tune navigation aids or maintain an eye-watering focus on the approach charts.

Regardless of whether procedures—including approaches, standard instrument departures (SIDs), standard terminal arrivals (STARs), charted visuals approaches, FMS visual approaches, or other published procedures—are selected by the crew or manually inserted waypoint by waypoint, the discipline is unchanged: the charts are then used primarily to verify the accuracy of the crew's programming of the FMS. When the procedure is briefed, it is done by the pilot flying, briefing off of the FMS and the appropriate displays. Once you master this skill, you will have noticeably closed the performance gap that results from briefing exclusively off the chart—with all its details—to focusing on the aircraft's performance and any redundant steps you have programmed to ensure 100 percent compliance with your clearance, including lateral and vertical navigation restrictions, which are constantly added to arrival and departure routes worldwide. This simple, three-step formula can be applied universally to instrument approaches, departures, arrivals, en route segments (such as oceanic or special route clearances), and descents. In doing so, you will be transforming your best preflight plans into action, communicating intent, and ensuring complete compatibility between what you intend to have happen and what the actual aircraft flight path does.

The Three-Step Glass Cockpit Briefing

1. "Build," load, or select the appropriate procedure from the FMS database with direct reference to the current chart.

2. Rigorously check the FMS programming for compatibility with the aircraft clearance.

3. The pilot flying (workload permitting) briefs the procedure from the FMS, while the pilot monitoring (ideally) checks the programming against the appropriate chart(s).

Crewmembers who want to achieve reliable, repeatable, expert-level performance apply this formula across innumerable tasks in advanced technology aircraft of all types; it is suitable for novices and experienced crewmembers alike, and it needs very little modification to work in almost any situation. You can practice this virtually anytime, anywhere, and you can become very proficient, very quickly. As simple and obvious as it sounds, it is one of the consistently present techniques that the best performers we have observed use to separate their performance from that of crews who proceed without the benefit of such simple, but effective rigor. (It is such a multiplier of crew performance that a number of large organizations and airlines have adopted it as SOP.) When you hear this kind of brief delivered effectively, you understand what Klein is referring to when he states, "The art of describing your intent is to give as little information as you can."[7]

Over a decade ago, when research into crew performance on the automated flight deck was beginning to peak, one study highlighted the substandard behavior of crews in briefing and verbalizing entries to the FMS. In his study of four U.S. Airlines, William E. Hines reported that ". . . crews in advanced technology aircraft who scored substandard on preflight briefings (10% of crews observed), 40% were also rated substandard on verbalizing FMC entries at the approach phase of flight, and 58% were rated as substandard in *overall crew effectiveness.*" Hines' data goes on to show that "Issues concerning automation usage permeated every phase of flight for all airlines, but became more pronounced during the cruise and approach phases of the flight . . . [with] a significant problem with crewmembers communicating and acknowledging changes in the FMC during all phases of flight."[8]

As you read the final words of the crew of Eight-Five Victor-Tango, it is difficult not to insert a desperate final plea, if it could be possible, before the approach is commenced, to request a holding pattern or delay vector so that the crew could have a second chance at briefing their approach to Houston-Hobby's runway 04:

0611:13/2,824 ft
FIRST OFFICER: "Localizer's alive."

0612:23/2,265 ft
CAPTAIN: "I can't get approach mode on my thing."

0613:14/1,757 ft
FIRST OFFICER: "We're high on the glideslope now."

0614:05/901 ft
CAPTAIN: "Did you change my frequency??"

0614:40/323 ft
FIRST OFFICER: "Coming up on two forty-four."

0614:47/213 ft
End of CVR recording

A briefing for this approach compatible with effective preflight planning and mission preparation would certainly have allowed this crew to leverage the technology in their favor, and very likely would have resulted in a different outcome. To begin to apply this skill to your own briefing patterns, simply view the ILS approach chart provided in Fig. 5.1 while verbalizing your intent according to the three-step briefing process above.

The Importance of a Comprehensive Debriefing

Corky Meyer, whose quote is on the masthead of this chapter, worked for Grumman Aerospace for over 35 years, test-flying aircraft from the F6F Hellcat to the F11F Tiger. His wisdom on the importance of briefing and debriefing every flight is as relevant on today's modern flight deck as it was when he was learning (very likely the hard way) the importance of both. It reminds us of the age-old adage that many professionals invoke when they refer to their own performance while steadily building on their personal expertise: "The best flight I've ever had is always my next one." It is an axiom of personal excellence that is hard to argue with, and we will show in the final pages of this chapter the vital role this plays in mastering skills associated with briefing and debriefing.

In a more contemporary reference, Gary Klein reminds us how experts become experts in the first place: "Experts are not just accumulating experience. People become experts by the lessons they draw from their experience, and by the sophistication of their mental models about how things work."[9] In a system that joins complex automation with human operators, Klein's observation is vital to an individual's professional progress. But what seems to be a simple concept eludes many crewmembers and organizations, and traditional debriefing strategies do not take into account the importance of analyzing unexpected and unanticipated "automation surprises." One of the greatest factors in reducing the time it can take to gain expert-level knowledge of automation is the skillful use of debriefings to enhance lessons learned, and to turn those lessons into cues that crewmembers can rely on in future, similar situations, so that unsafe or unpredictable conditions can be reduced and systematically eliminated.

In Chap. 3 we described the FMS in a way that places the crew in command of the information that is wrought from the variety of components, from power plants to global positioning systems. We touched on the logic of some of these systems, but stopped short of explaining how every type of FMS uses the information available to it, combined with inputs from the crew, to make and execute decisions which can have an immediate and dramatic (or delayed and subtle) impact on the aircraft's flight path. That would make this a multivolume series, much of which would need to be excluded by each reader depending on the type of FMS he or she is familiar with.

Information that is specific to individual aircraft and systems is available from the manufacturer's or the individual organization's operational flight documents—but there are some universal truths that are independent of specific systems. One is that automation will not always behave as we expect or anticipate. Another is that when it does not, not only do we have to intervene to correct flight path deviations, but also, as soon as it is practical, crews must assess the conditions that led to the unexpected situation, and debrief the experience for themselves (and if necessary, for their colleagues as well).

An invaluable resource for aviators of any level of experience is NASA's anonymous Aviation Safety Reporting System (ASRS) which allows crewmembers to report conditions or situations that led to potentially dangerous or unexpected situations which would otherwise go unreported (and without jeopardy to the reporting crew!). ASRS gathers these experiences by the hundreds every day, archives them, and offers a robust search function that is accessible to anyone with an internet connection. The system has been hailed as one of the best innovations in promoting flight safety ever devised. In just a few minutes, a search for situations involving crews of advanced transport aircraft whose actions resulted in missing an altitude constraint on an arrival procedure turned up more than 4000 results. The ASRS database exists to search events involving specific aircraft models (not just manufacturers), allowing individual pilots to examine thousands of situations that might apply to the equipment they operate specifically and to realize countless lessons learned by other crews operating nearly identical aircraft! Consider the following two events pulled from NASA's own ASRS, and you can appreciate the strong relationship between unexpected events on the flight deck and a timely debriefing that we advocate in this chapter.

Synopsis 1: B737-700 Captain reports descent below published altitude during transition from SEAVU fix to the ILS 24R at LAX. Deviation occurred after the second runway change during the arrival.

Narrative 1: While flying the SEAVU 2 Arrival to LAX, we were planning the 24R approach. We planned, briefed, and loaded the approach into the FMC. Approach gave us the arrival to 25L. We made the necessary changes and briefed and loaded the new approach. Then they changed the arrival back to 24R. We reloaded that approach. During the transition from the arrival to the approach, we crossed SEAVU fix at 10,000 ft and continued to descend to 9500 ft. We should have maintained 10,000 ft until crossing the next fix which was SKOLL. The First Officer was using the autothrottle and VNAV functions but didn't notice the descent below 10,000 ft as the workload went up during the runway changes. I was the PM but didn't catch the descent until we were below 10,000 ft. Both Pilots need to improve their proficiency using VNAV and autothrottle. We both thought it was going to maintain the descent profile. Even though the workload went up, I should ensure the aircraft was doing what we thought it was going to do.

Synopsis 2: A B737-700 crew expected the LAX Runway 24R SEAVU TWO but later were cleared for Runway 25L. After the FMC programming change the 280 KT KONZL crossing speed dropped out, unseen by the crew, causing the aircraft to slow and cross KONZL 400 FT high.

Narrative 2: During descent to LAX on the SEAVU TWO Arrival, due to a runway change as we approached KONZL (17,000 FT / 280 KIAS), we were several hundred feet high. We were expecting Runway 24R and then changed to Runway 25L. I was the pilot flying and I had, previously, changed the speeds on the descent page to cross KONZL at 280 KIAS. After the runway change, I suspect the descent speed reverted back to 261 KIAS. This caused the autopilot to command a level off to slow from 280 KIAS to 261 KIAS. We were on the proper descent profile, but with the level off about 3 NM from KONZL, by the time we assessed the situation, it was too late to push over at a reasonable rate. We ended up crossing KONZL at about 17,400 FT and 270 KIAS. The rest of the approach and landing were uneventful.[10]

It is clear in both narratives that the respective crews looked back at their flight conditions soon after the event occurred, carefully considered the factors that they felt contributed to the deviation, and resolved the action steps which would prevent similar excursions from recurring in the future. What is more, to cement their learning experience, they wrote an expository synopsis and reported it for the benefit of their colleagues worldwide. It is fair to say that these two crews will have little trouble in the future with altitude constraints along the SEAVU arrival to Los Angeles. We are left to wonder, however, what the format of their debriefing took, and how they proceeded to assess the crew actions, FMS logic, and workload issues that created the deviations. Was it a "crew bus debrief" on the way to the hotel, or was it a dedicated effort in the flight operations briefing room? In either case, to optimize the development of expert performance and multiply the benefit of these occurrences, we suggest the adoption of a simple formula, similar to the briefing that crews should follow after every event like the unexpected altitude deviations described above. Our formula is a modification of one that is advocated by our friend and colleague Tony Kern and is the best debriefing checklist we have ever seen—the Clear Signal Two Minute Debrief.[11]

1. *Safety:* These are things that must be discussed—was there any action that caused our safety margins to be diminished?

2. Standards: Was there any action by either (or both) crew members which violated established procedure and may have caused either or both to impact a line check or a flight?

3. Unanswered questions: Did anything occur that any member of the crew did not understand, caused by either crewmember action or unexpected action of the automation?

4. Improvement opportunities: Are there any areas in which we could do better because of knowledge we did not invoke, or knowledge which we have gained from this experience?

This protocol counts as the best we have ever seen because it not only provides the proper priority for debriefing virtually any flight, but also sets a constructive tone that contributes to the "mutual support" climate that is so vital in the advanced technology cockpit. Concerning the unexpected performance of the technology itself, the most valuable learning opportunity is created by assessing what each crewmember thought of the sequence of events along the flight path, and how they perceived and understood it. Remember, we are building expertise toward the goal of multiplying available experience by capturing what it is that the best do, how they do it, and how they know what it is that they are doing so well. Recognizing that a certain set of conditions is developing—or has the potential to develop—by processing the environmental cues (for example, airspace or clearance constraints, the performance limits of the FMS and flight guidance, a tailwind during descent) and understanding the factors involved is a hallmark of expert performance. In some domains, this is often categorized simply as "intuition," as if it were some kind of special power of experts.

Intuition is not magic. Two leaders in the field of decision making and expertise, Gary Klein and Daniel Kahneman, both endorse the same, concise definition of skilled intuition: "The situation has provided a cue: this cue has given the expert access to information stored in memory, and the information provides the answer. Intuition is nothing more and nothing less than recognition."[12] Being able to recognize a developing situation before it becomes hazardous is a front-line defense against bad outcomes that every aviator should constantly be developing. If, on your next flight or mission, you apply the tenets of this principle of Automation Airmanship, *briefing and debriefing*, then you will inevitably raise the level of performance of not only yourself, but your crew as well.

Checklist for Success: Briefing and Debriefing

☑ Briefings include critical operational information, but are not exhaustive in every detail: effective briefings do not contain "too much" information and thereby "weigh down" crewmembers with unimportant information and/or data.

☑ Brief Standard Instrument Departures (SIDs)/Standard Arrivals (STARs) and approaches with reference to the FMS or equivalent system, if available, and compare to map or other plan views.

☑ Brief intended autoflight mode configurations for critical phases of flight.

☑ During low-workload phases of flight or as soon as practicable after the flight, debrief unexpected/unanticipated automation behavior and incorrect setup which have occurred during the flight or mission.

☑ Conduct a formal postflight/postmission debriefing that addresses safety, standards, unanswered questions, and improvement opportunities.

Notes and References

1. Meyer, Corwin H. *Corky Meyer's Flight Journal: A Test Pilot's Tales of Dodging Disasters – Just in Time*. Specialty Press, North Branch, MN. 2006. p. 53.
2. National Transportation Safety Board Accident Report N85VT.
3. Gary Klein, *Sources of Power*, MIT Press, Cambridge, Mass., 1998, p. 219.
4. Ibid., p. 222.
5. Ibid., p. 225.
6. Ibid., p 229.
7. Ibid.
8. W. Hines, "Flight Crew Performance in Standard and Advanced Technology Aircraft," *Proceedings of the Ninth International Symposium on Aviation Psychology*, The Ohio State University, Columbus, 1988, pp. 1066–1072.
9. Gary Klein, *Streetlights and Shadows: Searching for the Keys to Adaptive Decision Making*, MIT Press, Cambridge, Mass., 2009, p. 101.
10. NASA ASRS database search for FMS equipped transport aircraft + Crossing Restriction Missed.
11. The Clear Signal debriefing is attributed to Dr. Tony Kern and the Blue Threat training program.
12. Daniel Kahneman and Gary Klein, "Conditions for Intuitive Expertise, a Failure to Disagree," *American Psychologist*, September 2009, p. 520.

CHAPTER **6**

The Third Principle: Data Entry

You must learn to place less value on all that you can remember and more on those few things that you must never forget.
—Marcus Buckinham, 2005[1]

On March 20, 2009, in Melbourne, Australia, a veteran and experienced crew arrived at their aircraft, an Airbus Industrie A340–500, for a routine, scheduled passenger flight to Dubai, United Arab Emirates. In the course of their duties to prepare the aircraft and flight deck for departure, a critically important setup step was incorrectly performed and subsequently went unnoticed until, at the calculated rotation speed on the runway, the aircraft failed to achieve a normal liftoff attitude and become airborne. During the critical few seconds that followed—as the aircraft struggled to lift off the runway—the crew took decisive action to avert what could have been a devastating and tragic accident. The aircraft (and some airport facilities on the ground) sustained significant damage, yet the incident slipped quietly into the annals of "near disasters" with little fanfare across the industry; only a few safety organizations and industry journals would report on the occurrence and its significance in the following weeks and months. On board were 257 passengers, 14 members of the cabin crew, and 4 cockpit crewmembers.

At the root of this incident, since categorized by international safety organizations as a "near loss," was the inadvertent entering of the wrong takeoff weight into the aircraft's portable flight planning computer by the first officer during preflight. Cross-checks which were designed to capture this kind of error did not function as designed, and the takeoff speeds and engine thrust settings which the crew would then enter into the FMS were grossly insufficient to allow for a normal takeoff (the magnitude of the mistake was on the order of 100 metric tons, approximately 220,000 lb).

In its investigation into the incident, the Australian Transport Safety Bureau (ATSB), that country's leading public institution for transport safety, would discover that the occurrence was neither isolated to the A340 nor unique, as they revealed ". . . similar take-off performance-related incidents and accidents across a range of aircraft types, locations and operators around the world."[2] Their concern would catalyze a broader investigation into this class of accidents, which has subsequently raised concern around the industry. Indeed, within nine months of the occurrence in Melbourne and despite the notoriety of this incident among operators of the A340, a second and similar near-accident at London Heathrow bore eerily similar details, as well as related events involving B767 and A330 aircraft, among others, in recent years.[3] The crews of these aircraft, like the crew in Melbourne, would each fail to accurately enter and subsequently cross-check critical flight data and, absent near heroic pilot intervention, narrowly avoid catastrophe.

In earlier chapters we have established, painstakingly, the notion that automation is a dutiful, tirelessly cool, and reliable resource for modern aviation. We have also touched on the main theme of this chapter, and the basis for formulating a discreet component of Automation Airmanship that explains the relationship between the pilots and crew, by now known as the familiar *wetware*, and the hardware and software that function so reliably at their fingertips. How can crews arrive at their aircraft rested and motivated, committed and dedicated, experienced and competent, yet miss critical details, even when these details are integral to established protocol, procedure, and SOP? Not only will we answer this question, but also we will provide a clear path through distractions that potentially plague even the safest of operations.

Errors and cross-verification failures are not limited to preflight, but find their way into the actions and omissions of crews from preflight to postflight in every variety of aircraft, across every culture, at all levels of technology, and in every corner of aviation. Perhaps the most prominent accident in this category is one that has been written about and analyzed so thoroughly that it has become a virtual cliché in the circles of contemporary aviation safety literature: the official report of American Airlines flight 965 in Cali, Columbia, in December 1995 can be obtained from numerous official (and countless unofficial) websites (a Google search of the same will produce almost 10,000 results). One of the best accounts of how an incorrectly entered piece of navigation data would lead to the total loss of a B757 and nearly every passenger onboard the aircraft that day is related in the 2010 book by William Langewiesche *Fly by Wire*, in which the now well-distilled episode is recounted elegantly in the context of automation nearly 15 years later. Nonetheless, what is lacking in many of the accidents and incidents is the presence of a repeatable, reliable, deliberate, and rigorous discipline of flight deck data management; if accounted for and required in official procedures, in virtually every accident involving automation, these procedures are often violated. The very best-performing pilots and crew have been practicing elements of this essential skill for decades, and no reasonable aviator would dispute the need for, or role of, this requirement; we intend to provide both the background "systems knowledge" of the human component, introduce a proven expert formula for flawless performance, and merge this with companion Automation Airmanship principles to invigorate organizations' and individual crewmembers' defenses. In turn, crews will be able to deploy this knowledge in the face of the relentless avalanche of information that is constantly streaming onto the flight deck (and into mission compartments of specialized aircraft), begging for the immediate and total attention of the crew, without pause and without prejudice. As we plunge into the basic condition facing expert and inexperienced crews alike, it is worth a moment of reflection excerpted from Langewiesche's more contemplative volume, *Inside the Sky*, published in 1998: "Pilots have to take their fate firmly into their own hands. Airplanes speak to them through their controls, but they remain inanimate machines and mindlessly unforgiving."[4]

First, the Wetware

Advances in cognitive psychology, cognitive neuroscience, psychology, brain imaging, learning, and other related disciplines over the past half-century have provided numerous popular notions of how humans function in high-risk, high-reliability organizations that are frequently attended by stress, time constraints, and other factors stretching the human system's capability. You may read this and other parts of this book through the filter of commonly accepted notions of thinking, reasoning, and other activities of the mind which have been formed by formal or informal exposure to the knowledge of the human brain and how it functions. In this chapter we aim to bring forth the most relevant and contemporary information on the topic of mental capacity as it relates to managing information and data on the modern flight deck (and maybe inspire a little personal research into the subject on your own). For this, we consider the extraordinary power—and limitations—of the human brain.

In comparison to any other known system, "the human brain is the best organized, most functional three pounds of matter in the known universe."[5] With respect to the domain of aviation alone, it is responsible for the Wright Flyer and the Concorde, the Spirit of St. Louis and the Space Program, the Berlin Airlift and the Battle of Britain, Super Carriers and the Denver International Airport, the Sperry Gyrostabilizer and the increasingly powerful flight management system (FMS), among numerous other accomplishments. As many experts tell us, and most all of us know, it is also responsible for approximately 20 percent of all aviation accidents.

We know that our brains can contemplate a wide range of information, memories, thoughts, and ideas; in fact, in our own industry we have created a working environment (high altitude, long range, often severe weather, and even the vacuum of space) in which our own brain cannot survive for more than a few seconds outside of the protection provided by the aircraft hull itself. In spite of its vast capabilities, it is also a very limited system. On the flight deck (and mission compartments of specialized aircraft) our brain combines with the sensory receptors of the rest of the carbon-based unit to, as Robert Sylwester remarks, "monitor the surrounding environment for specific changes." The ranges in which these systems, including the brain, operate are

> . . . about 130 degrees on the Fahrenheit scale, 30 or so odor-related molecules, 10 sound octaves, 4 food properties, and the narrow band of visible light in the broad electromagnetic spectrum. These changes in the surrounding environment may not seem like much, but they are the primary source of all the information our brain gets about what's happening outside our skull. . . . Our sensor system is genetically quite limited.[6]

Knowing more about these limits, as they apply to critical functions such as data entry, cross-checking, and procedural steps designed to capture and eliminate errors, is a hallmark of experts who work in our domain. To this end, we examine what many experts know about memory and managing information accurately, within the limits of their own brains.

Most of what we are tasked with accomplishing on the flight deck is comprised of predictable routines that in most cases are also maintained by policy and procedures, including checklists and even established "techniques." Yet every flight or mission is different from every other in any variety of ways (weather, equipment status, crewmember proficiency, fatigue, etc.), requiring even the most robust procedures to be flexible enough to handle conditions that are frequently shifting and often (as in the case of preflight) are accompanied by time constraints. Marvin Minsky, cofounder of the artificial intelligence laboratory at MIT and a pioneer in the area of artificial intelligence, summarizes what our brains are doing while we are engaged in activities that require strictly adhering to procedure while also accommodating other demands in a highly dynamic environment: "In order for a mind to think, it has to juggle fragments of its mental states."[7] In his research and writing on the topic of what is commonly known as "multitasking" in popular culture, Minsky offers a brilliant sample of what this requires (we are taking some license with his approach by fitting it into the context of the glass cockpit flight deck). If you can, visualize the final minutes of preflight, just prior to closing all aircraft doors and starting engines (a scenario you may be very familiar with). You are shifting your attention back and forth from a variety of demands including the gate agent or purser, mission crew, maintenance personnel, FMS, and other setup steps, among many other "agents" in the system. At each interaction, you form ideas and images in your mind, and some interrupt and overlap others; you consider alternatives as you learn that equipment has become inoperative, or suddenly gets repaired just in time. Your fellow crewmembers interrupt you with vital flight information, and your data link (which knows nothing about social protocol) rings with urgency, pushing other competitors for your attention out of its way. In one instant, you are concentrating on a very small piece of critical information, and in the next instant you form thoughts surrounding the entire flight or mission (or even the extended trip or mission over days or weeks). How do we manage to do this "juggling" in the face of so many distractions? For this answer, we look to the research and conclusions of one of the world's leading cognitive scientists, Joseph LeDoux. "The answer," LeDoux continues, "is that the mind uses something called working memory."[8]

How many times have you read back a radio frequency to air traffic control during a frequency change, only to forget it seconds later after being momentarily distracted by other traffic coming into view through the windscreen? LeDoux explains that this happens because you put the information into working memory, a "mental workshop" that accommodates only one task at a time. As soon as a new piece of information or task engages your brain's working memory, the old information is pushed out. Unless you write down the information, rehearse it, and manage to resist other distractions that are vying for your attention, the information is *gone*. LeDoux goes on to describe the prominent role of working memory in all aspects of thinking and problem solving—from the simplest things (such as changing radio frequencies) to coordinating the countless details required of complex tasks (such as operating a modern aircraft across oceans and continents).

This chapter is about how advances in brain science can equip you with greater defenses against errors when handling critical flight data. Take the following conclusions from the research community and apply it to your own experience: ". . . working memory is not a pure product of the here and now. It also depends on what we know and what kinds of experiences we have had in the past. In other words, it depends on long-term memory."[9] How you handle information in planning and programming critical aircraft systems is a combination of your skill in blending efficient use of your working memory with the knowledge your brain has already stored in long-term memory. Both working memory and long-term memory, cooperatively, combine to guide our actions and decisions. Most of the time this is good; sometimes, the fallibility of this system can lead to errors that can have disastrous outcomes. If we commit an error as a result of a fault with our working memory (transposing, perhaps, a single number while entering the takeoff weight into the performance software, resulting in an error with a magnitude of 100 metric tons (220,000 lb)), the network of experiences contained within our long-term memory (including what we know about defenses such as policies and procedures involving cross-checking and verification) should provide the opportunity to catch the mistake.

When you are actively and consciously applying your working memory, the "magic number" of things that can be active in working memory is seven (for some, this number can be as high as nine, for others as few as five). *This is a limitation that is worthy of committing to long-term memory.* (Do it now.) Do not confuse it with short-term memory; it is more than an area of temporary storage, according to Minsky, LeDoux, and other leading scientists. It underlies our ability to form thoughts and ideas, and take action, at the higher level, the level that the brain researchers have described as "executive function." A word picture that may help to describe the executive function is that of a computer, which moves information from permanent memory (ROM) to a CPU, with active memory (RAM) scheduling tasks to be performed using the active memory.[10] Extensive, detailed, and valuable information on how we think, reason, and decide can be found in LeDoux's book *Synaptic Self* and in his earlier works as well. Here is a summary of the executive function, again with some liberties taken, which applies the findings directly to the flight deck:

1. Through executive functions, specialized systems are directed to attend to certain specific stimuli and to ignore others, depending on what working memory is engaged in.

   You miss a traffic call from ATC because you were listening to ATIS for the destination airport.

2. Executive functions plan the sequence of mental steps and schedule the participation of the different activities, switching the focus of attention between activities as necessary.

   You brief your departure, engine startup to the top of climb, and compile information gathered from dozens of sources to describe the departure from a large congested airport and out of a busy terminal environment.

3. Executive functions are crucial to decision making, allowing you to choose between different courses of action given what is happening in the present, what you know about such situations, and what you can expect to happen if you do different things in this particular situation.

   You properly assess an abnormality or emergency which involves a loss of critical pitot-static air data, and in choosing the correct procedure isolate a potentially hazardous source of inaccurate information for the entire duration of the flight or mission.

4. Executive function is not all-powerful; it basically can do one thing, or at most a few things, at a time (multitasking as popularly referred to is a myth that most people hang onto at their own peril).

   Policies and procedures do not fade from importance as experience is gained in the field; they remain invaluable tools to back up the wetware, and should be treated as inviolable: no matter how many times you have taxied to the same departure runway, you always call for the checklist to be read by the first officer or copilot.

5. The executive function may appear to be doing many things at one time when scheduling the sequence of steps in a complex task, but the things are related to the overall goal.

   "Harmless" chatter about nonoperational topics while flying a nonprecision approach at night, for example, fouls the ability to plan the executive function, no matter how familiar the crew may be with the procedure.

6. If the executive has to work on multiple unrelated goals at the same time, the system begins to fall apart, especially if the goals conflict with one another.

   Allowing multiple, unrelated "agents" to interfere with briefings, planning, programming, and cockpit setup is a sure way to put pressure on the executive, and risk its breakdown.

The science of long- and short-term memory, working memory, and executive function has made remarkable gains in recent decades. We now know what we can and cannot do (lots of party tricks have grown up around this knowledge in recent years), and we can make decisions about how we engage the avalanche of information that is generated aboard modern aircraft. Dan Schacter has emerged as a leading researcher, and in his book *How the Mind Forgets and Remembers: The Seven Sins of Memory*, he cites the risks associated with relying on memory and how they can impact our lives.[11]

1. *Transience:* the inability to hold on to information.

2. *Absent-mindedness:* our annoying capacity to fail to pay attention to what we are doing.

3. *Blocking:* the failure to pull out that fact or name that is on the tip of your tongue.

4. *Misattribution:* typified by the belief that a memory was formed in one situation when it actually occurred in another.

5. *Suggestibility:* memories implanted by outside influences like media, therapy and cultural norms.

6. *Bias:* which leads us to revise our memory of a situation to make it fit what we feel or think now, long removed from when the memory was created.

7. *Persistence:* emotional arousal, which makes any memory stronger, can be good, but can also be debilitating if the memory is of a traumatic nature.

Memory and executive function factor heavily in performance on the flight deck, and expert performers rely on their knowledge of their individual capability to manage flight deck data reliably, and to skillfully involve the support system of policy and procedure without hesitation, to ensure that they are not relying only on their own (limited) ability to juggle multiple tasks. The very best do a few other things as well.

Expert Data Entry Protocols

Data entry errors fit near the top of the stack of human performance errors that arrived with the first automated aircraft and persist today. In an environment where errant information can lay seemingly quiet and disguised until its time arrives to impact the flight or mission, it seems almost certain that mistakes will continue to cause accidents as long as humans sit atop the decision-making hierarchy. We do not think that this dystopian view has to be accepted; nor do we believe individuals and organizations need to adopt the viewpoint that some accidents will inevitably occur, as part of the "cost of doing business." Quite the contrary, we think that performance will be driven up and error rates will reverse, and we have a generation of expertise that can act as beacon of action for the rest of us.

In 2000, Wesley Olson and Nadine Sarter published research on the difficulty that pilots can have detecting keystroke errors that might not have an impact for many minutes (an incorrectly entered latitude or longitude on a long flight plan waypoint string in the FMS, or perhaps an incorrectly inserted non-directional beacon (NDB)).[12] They term the act of entering configuration data into the FMS as actually making (reversible) commitments about the flight path of the aircraft. Without knowledge of the logic, they argue, pilots and crew are less likely to notice their errors. As individual crewmembers increase their experience and knowledge of the logic of the FMS, incorrect or inaccurate "commitments" actually become easier to spot, and crews are more likely to act to "reverse" the commitment when found. It happens every hour, thousands of times, on flight decks around the globe. We have watched many experts (and some organizations have even committed these habits to policy and procedure) adopt excellent strategies that leverage the knowledge of both human limitations and that gained from the research community. The controlling insights which guide them through this process and are counted among those things that experts know they can *never* forget, are arguably the most telling signs of a high level of automation airmanship and expertise.

Let us take the data entry habits of experts one at a time, starting with recognizing FMS data input as a specific flight deck activity. Experts know the added risk to the operation of combining data entry activities when the aircraft is moving on the ground: during initial setup inputs to the FMS (and other systems requiring input such as performance computers, targeting systems, and electronic chart systems), high-performing, experienced crews always do data entry when the aircraft is parked and they are executing no other procedure or checklist. When in flight, only the pilot monitoring (sometimes referred to as the "nonflying pilot"), or PM, is allowed to make FMS inputs. In rare instances, during low-workload cruise segments, some organizations allow the pilot flying (sometimes called the "handling pilot"), or PF, to make FMS inputs, but only when advising the PM that he or she is doing so. These practices have one outcome in mind, and that is to promote the concept that the time spent inputting data into the FMS is *inviolate* and must be held up as a *critical flight activity* by individual crewmembers and the organization.

For nearly a century, the concept of *mutual redundancy* has been a foundational CRM principle in all aircraft that require more than one pilot in the cockpit. The evolution of the respective roles of the pilot-in-command and the second-in-command has been ongoing since the earliest days of aviation, with many memorable tracts written and movies made depicting the classic "captain" and "copilot" and the relationship that holds the flight deck team together. With many flight departments, airlines, military flying organizations, and even most manufacturers adopting the terms *pilot monitoring* and *pilot flying* designations in aircraft flight manuals (AFMs), flight crew operating manuals (FCOMs), company flight manuals (CFMs), SOPs, and checklists, the classic model of the steely-eyed, all-knowing captain and his trustworthy journeyman copilot are true vestiges of a bygone era. The impact of even the smallest input error on the advanced flight deck has made it necessary that all crewmembers operating these aircraft view themselves equally as "steely-eyed *glass-cockpit pilots*." Since the captain cannot be expected at all times to make all the critical FMS inputs, he or she must be able to rely on fellow crewmembers to input information as expertly as if the captain had done so personally.

The mutual redundancy provided by clear procedures for verifying data that have been input to the FMS is, in its purest sense, a commitment between both pilots that no setup error will be tolerated. These pilots adhere strictly to any procedure or SOP that requires them to verify data against the flight plan, performance computer, or other backup resource. This principle can be extended beyond required verification checks and applied to in-flight programming with the same strict requirements required to fully realize the term *mutual redundancy*. Here is a simple example of this procedure in action:

"I just put in the holding pattern on the arrival—will you check it before we get there to be sure I got it correct?"

The last essential component of this concept is that at *no time* on the flight deck is "secret typing" allowed to occur. Making inputs to the FMS, in most cases, has a direct influence on the current or future flight path of the aircraft and, as such, must be coordinated between crewmembers as with any other procedure affecting the flight path of the aircraft. Again, this is a pledge of excellence between crewmembers that there will be *no* unexpected surprises as a result of independent data entry. Once again, here is an example of how this works through clear communication:

> "I'm going to work ahead in the flight plan and add the 'expect' altitude constraints along the arrival procedure. . . . It sounds like the guys ahead of us are all being given the same crossing restriction by ATC. . . ."

As we will see in upcoming chapters, the fact that there is more communicating, and most of it is more specific, is a distinct feature of expert performance on the glass cockpit flight deck.

Aphorisms abound surrounding the concept that when things are going extremely smoothly, we should be extra vigilant in checking our environment for threats. Although there is little science on the subject, it is a commonly held belief with which most of us can identify. In his book *Inviting Disaster: Lessons from the Edge of Technology*, James Chiles stretches a classic line from Shakespeare's *Othello*, substituting setup error in complex systems for the original reference to clandestine homicide. Chiles' version states that disastrous outcomes often have an element in which the highly trained and reliable professional, through familiarity and routine, is guilty of ". . . working not wisely, but too well."[13] The similarity resonates with anyone who has achieved a level of proficiency with the technology she or he operates, such as to be capable of performing it rapidly and with little conscious thought. *Familiarity and routine are not a substitute for proficiency and vigilance.*

In their transition to glass cockpits, pilots who have flown less sophisticated aircraft inevitably become overwhelmed by the seemingly hundreds of "pages" contained in the FMS, many with their own embedded pages as well. The deeper you drill into the capability of your own FMS, it seems, the more there is to remember. *There is no shortcut to learning the basic organization and then building on that knowledge through practice and experience.* The very best crews have demonstrated repeatedly that they are able to quickly find the right page on their FMS for data access or input through expert-level knowledge of their FMS menu and, in turn, knowledge of the actual pages themselves. Usually the second wave that hits the shore of the unexpected and unprepared is the notion of how much reading of displays is required and how much detail is contained in FMS messages, ECAM, CAWS, EICAS, ACAWS, or other automated alerting system, as well as the EFIS display. It has been said by many expert glass cockpit pilots that every one of these aircraft is "a reading pilot's airplane" (note that this rule does not say "speed-reading"). A corollary to the reading pilot's airplane concept is the equally important requirement to *read what you just typed for accuracy and correctness* before pushing the insert or execute prompt. What have these successful, veteran pilots learned through years of experience that can be decoded and taught to others? With respect to data input, they take the time to read their FMS scratch pad before they insert or execute information into the FMS. This sounds simple, and many veteran instructor pilots and evaluators will tell you that the quickest and nimblest fingers make the most mistakes—simply by not by taking the time to read their work—*working not wisely, but too well*. It is the one quality control step that we can implement at the personal level that can have an immediate and positive impact on overall data entry effectiveness.

Similarly, experts are tireless at cross-checking and verifying data—*all data*. Very few inputs into the FMS can be made without verification by the other pilot in the cockpit; most require some form of cross-checking. Checking the work of others is a natural complement to rigorously checking your own. More than likely, you can immediately recall, with little effort, the last time a crew you were a part of caught an "input error" during cross-checking. Here is a short list of critical items that need to be cross-checked by both crewmembers following input into the airplane's FMS or flight guidance system:

1. Altitudes, including those programmed as waypoint constraints
2. Speeds, whether input through the FCP/GCP/MCP or into the FMS
3. Nav data (all routes, SIDS, STARS, approach procedures, and modifications to the aircraft lateral and vertical flight path)
4. Altimeter settings
5. Performance data (those that are entered manually into the FMS or other interfaces with automatic control systems)

Most organizations have specific, detailed procedures for each of the above data; those that do not, should.

Many prominent, recent disasters share this same element: information that was known to some but not shared with others, which played a major role in the accident. Notable are the *Challenger* and *Columbia* shuttle disasters, the grounding of the Exxon *Valdez*, and the explosion and sinking of BP's Deepwater Horizon oil rig. On the glass cockpit flight deck, there are in fact design features that can actually obstruct the sharing of critical flight data: each pilot (and on some specialized aircraft, mission crewmembers) has an individual

FMS input unit at her or his crew position, making it difficult for one pilot to see the input actions of the other. Many flight deck designs have a variety of methods, including the FMS, to input the same data, making it easy to unknowingly and unintentionally conceal or disguise data input. If these structural impediments to smooth information exchange and cross-checking are not enough, we know that what seems obvious to the pilot making data inputs can be completely unnoticed by the pilot flying, who is (and should be) occupied with other flight-critical tasks. At this point you should be recalling the "power to read minds" concept that we developed earlier: you can apply that skill to solving this problem and perform at the same level as the experts. Simply communicate more often, and more clearly, with more specific information than you may be used to doing previously (a concept we develop extensively in Chap. 7). In conventional aircraft that many pilots flying glass cockpit aircraft were trained in or operated for long periods of time, it was easy to see and verify the dialing-in of the ILS final approach course using the OBS knob on the HSI.[14] Likewise, since much of our FMS programming can be data that are not instantly recognizable on the situation displays in the cockpit (such as wind data and some performance data), it is important that the crewmember making the input clearly brief those actions with the other pilot—we all know that surprise on the flight deck is something that all experts seek to avoid.

Most of us can identify with the clutter that can accumulate on the various displays on the flight deck through the actions of pilots programming information into the FMS, as well as with the selection of the variety of additional displays that are designed to help maintain the highest level of situational and mode awareness. Think back to your first few hours of training in advanced aircraft and how the amount of information displayed seemed overwhelming. You learned to focus on the essential information, disregarding that which was (at the time) unimportant. A very important aspect of managing the data on the automated flight deck is to ensure that the displays are kept up to date and uncluttered and that map displays are set at the appropriate scales. You may also want to consider which specific pages you want to have displayed on the respective FMS control units. For example, it is always convenient to have the PM on the FMS page that gives instant access to the flight plan in the event of a route change or direct clearance after takeoff or during climb. Likewise, the PF may want to have takeoff performance or engine-out pages displayed for reference in the unlikely event of an engine failure on or after takeoff. Similarly, aircraft equipped with enhanced ground proximity warning systems (EGPWS), TCAS, predictive wind shear (PWS), heads-up displays (HUDs), or a variety of other enhanced information systems offer the choice to have *different* information displayed on each side of the cockpit. The amount of information available for display in the glass cockpit long ago surpassed the capacity of the wetware to process it quickly and accurately: it is up to the crew to exert a conscious effort to prioritize in order to realize the advantage that so much information can bring.

Reading the accident reports of dozens of accidents will eventually lead most aviators to conclude that technology in the cockpit can be one of the greatest sources of distraction on the flight deck. The input screens that pilots spend so much time looking at have been called a number of different things as they have become more common. Our favorite term of endearment is the *eyeball sucker*, a term that vividly describes how hard it is for both pilots not to look at the FMS input screen as information is being read from or input into the FMS. We have talked about other components of flight deck data management that will help to prevent both pilots from being "heads down" at the same time, such as "no secret typing" and achieving proficiency with the FMS, making data input more efficient and accurate. Some organizations even specify a required "head down" call when a pilot anticipates being occupied with FMS programming to the exclusion of other cross-checking duties for more than just a few moments. However you handle this important aspect of data management on the flight deck, be sure that you communicate with the other pilot (and crew, if applicable) when you notice any crewmember becoming fixated on the FMS, while at least one pilot should always be "head up" in the cockpit.

When we speak with groups across aviation and other industries, we frequently begin by asking the questions, "Is your job more difficult, demanding, and complex than it was 10 years ago? How about 5 years ago? Today? What do you think the future holds for your chosen field of professional performance, say, one year from today? Five years from today?" The results are overwhelmingly similar, across industries and across international and regional borders: working as front-line professionals in high-risk, high-reliability organizations, often with the assistance of evolving technology, continues to demand greater knowledge, dedication, and commitment as careers mature and technology proliferates. We often follow up these questions with one that is simpler and that causes many of the individuals to reflect deeply on their own commitment, professionalism, and vision of their own future: "What, then, are you doing today to ensure that you will, at a minimum, keep pace with the demands so that someday, in a moment that may define your professional legacy (and perhaps your organization's, too), you can meet that moment with the highest level of readiness?"

As the final chapter of Part II, Organizing the Advanced Flight Deck for Optimum Crew Performance, we have developed three major principles described by specific, observable, and learnable behaviors that experts perform during the normal course of their duties on the flight decks of advanced aircraft. You have, by now, begun to see the relationship between each principle and areas where they overlap or even merge to get the job of expert performance done. If you stop here and never read another page of this book, yet incorporate these principles into the way you manage your own and your crew's performance in whatever aircraft you fly, you will be better equipped for the rapid change that a demanding future holds. Part Two, after all, is all about preparation. If you continue to read the rest of this book, you will find knowledge and skills that you can use immediately, to complement what you have learned already and raise your performance even higher.

Checklist for Success: Data Entry

☑ Crewmembers locate FMS pages with ease, without "searching," and are proficient in entering flight data through each input pathway available.

☑ Data entry errors are immediately recognized and quickly resolved.

☑ Flight-critical data entries are cross-checked at a compliance rate of 100 percent.

☑ No "secret typing" or other input method is practiced by any member of the crew—entries are made with full knowledge of supporting crewmember(s).

☑ All displays are kept up to date, accurate, and consistent with crew briefings for phase of flight.

☑ "Head-down" time is minimized by both pilots, and in no instance during a critical phase of flight or taxiing are both pilots simultaneously preoccupied with data entry or verification.

Notes and References

1. Marcus Buckingham, *The One Thing You Need to Know*, Free Press, New York, NY, 2205, p. 26.
2. ATSB Media Release 2009/21, Tail-strike Melbourne Airport, VIC. 20 March 2009 A6-ERG airbus A340-541, December 18, 2009.
3. David Kaminski-Morrow, "Virgin A340 Take-off Miscalculation Defeated 'Robust' Checks," *Flight Global News*. 8 July 2010. London, UK. http://www.flightglobal.com/news/articles/virgin-a340-take-off-miscalculation-defeated-robust-checks-344187/
4. William Langewiesche, *Inside the Sky: A Meditation on Flight*, Random House, New York, 1998, p. 34.
5. Robert Sylwester, *A Celebration of Neurons, An Educator's Guide to the Human Brain*, ASCD, Alexandria, Va., 1995, p. 1.
6. Ibid., p. 57.
7. Marvin Minsky, *The Society of Mind*, Simon & Schuster, New York, 1985, p. 153.
8. Joseph LeDoux, *Synaptic Self*, Viking, New York, NY. 2002, p. 175.

9. Ibid., p. 176.
10. Ibid., p. 178.
11. Daniel L. Schacter, *How the Mind Forgets and Remembers: The Seven Sins of Memory*, Houghton Mifflin, London, 2001, pp 1–12.
12. Wesley A. Olson and Nadine B. Sarter, "Automation Management Strategies: Pilot Preferences and Operational Experiences," *International Journal of Aviation Psychology*, 10(4): 327–341, 2000.
13. James R. Chiles, *Inviting Disaster: Lessons from the Edge of Technology*, Harper Collins, New York, 2001, pp. 158–159.
14. The omni bearing selector (OBS) knob is a component of the horizontal situation indicator (HSI), commonly installed in older, "round-dial" cockpits that perform some of the basic functions of the primary flight display (PFD) in glass cockpit aircraft. To input the final approach course for an instrument approach in a round-dial cockpit, the pilot would reach up to the instrument panel where the HSI is located, turn the OBS knob to the final course (displayed on the instrument as a result), and usually accomplish this during the approach briefing. It's hardly an action that can be missed by the pilot monitoring or any other crewmember on the flight deck, making verification much easier than in many glass cockpit aircraft where the input is made through keypads away from the PFD.

Part

Harmonizing Essential Crew Capabilities with Flight Deck Automation

People in professions marked by standard methods, clear feedback, and direct consequences of error appear to appreciate the boundaries of their expertise. These experts know more knowledgeable experts exist.
—Daniel Kahneman and Gary Klein, 2009[1]

CHAPTER 7

The Fourth Principle: Communicating

The concept of intent can be applied to equipment as well as to people.
—Gary Klein, 1998[2]

At the beginning of this book we made a commitment to you that we would not spend our time on what has become known as basic crew resource management (CRM), but would develop concepts that reach higher into the domain of expert performance. For decades crew communication has been taught as a building block of CRM, and we expect that you already have a sound understanding of these basic concepts. Yet we also feel that, particularly in advanced aircraft, communication plays a more important role than ever before, and that this skill must be more thoroughly understood so that glass cockpit flight crews can better leverage their own performance and the increased capability of the aircraft they operate. In this chapter we examine the unique qualities of the communication that takes place in advanced aircraft between the human operator and the automated systems and that occurs between crewmembers who must use the systems to accomplish their shared goal of safe, efficient mission completion.

The modern aircraft cockpit, whether crewed by one pilot or a team of crewmembers, is one of the most complex workspaces in the world, with few rivals across other high-risk industries. The environment created by technology, mission requirements, individual ability, and outside influences (such as weather, obstacles, and terrain, and, in military and law enforcement applications, enemy offensive and defensive threats) values the effectiveness and accuracy of communications disproportionately to many other professional settings. For not only does the crew face the challenge of organizing and sharing vast amounts of information with one another, but also they must organize and share information with the automated systems, which requires them to provide *perfectly accurate* information to the aircraft, while also constantly questioning and resolving accuracy issues of the information generated by the aircraft systems and communicated *back to the crew.* Before we examine the processes that experienced crews use to communicate effectively, we look at how the automated flight deck can be organized to create effective communications between the crew and the aircraft's automated systems. We begin with a further explanation of the capabilities of both systems—the human operator and the automation.

Over the past few decades, the seemingly rapid advancement of information technology, fueled by the popular media and Hollywood, has created a sense that computers will soon have the capability to control much of our daily lives as devices transition from mere storage and processing of information (providing output to users within its environment) to learning from the environment itself *in addition* to being a powerful processor and storage device. (In aviation we are constantly reminded of the effectiveness of unmanned aircraft and the encroachment of these systems toward ground held for over a century by flight crews.) Artificial Intelligence (AI), a fascinating and rapidly developing field, is expanding daily on the power of exciting new research and holds much promise for high-risk/high-reliability domains such as aviation. One day AI will unquestionably have an increasingly important part to play in our industry. *But do not mistake the powerful systems of modern aircraft with having powers of intelligence greater than a well-trained professional flight crew.* To improve the way we communicate with the technology on the flight deck, we must understand what the technology is actually good at and recognize how the human operator complements this capability with what he or she is good at as well. Where these two vitally important components of our industry intersect is precisely where we find expert performers communicating effectively with both the technology and one another.

The computational capability of onboard aircraft systems is growing at an amazing rate; some of the technology assistants that we now have in the cockpit were not even dreamed of a decade ago, and the race to fill up the limited space on the flight deck and in mission compartments has been constant for years. On many advanced aircraft, the available equipment, displays, and controls have overtaken the available space in the cockpit in which to put it all. Yet new technologies proliferate on the flight deck as processing capabilities improve, memory is expanded and made increasingly inexpensive, and architectures are adapted to host more and more functions to process and communicate information to the flight crew. Even so, the mammalian brain remains stodgy when faced with the apparent need to adapt its own structure to keep pace with technology, incapable of expanding the storage and processing capacity over that of even our most distant ancestors. If you were to believe the pundits, it seems a losing battle for flight crews—that "fight between rivals," human and machine. In spite of Hollywood, which heralds the arrival of unmanned systems and pilotless transport aircraft, we confidently tell the steady stream of newcomers to advanced aircraft, "You are smarter than the aircraft you fly—considerably."

Even when you are tired, dehydrated, in the midst of an urgent in-flight emergency with elevated levels of adrenaline and cortisol coursing through your bloodstream, you are much more intelligent than your complex aircraft. It has been designed and built to execute tasks along very narrow, specific, and inflexible pathways. It is not, as AI researchers will attest, a ". . . general-purpose intelligence that can be set loose on any problem; one that can adapt to a new environment without having to be retrained constantly; one that can tease the single significant morsel out of a gluttonous banquet of information the way we humans have evolved to do over millions of years."[3] While the computational neuroscientists (those professionals at the center of AI research) race to create a machine that can respond to the level of humans, we remain the single component in the cockpit that is best suited for "intuition, pattern recognition, improvisation, and the ability to negotiate ambiguity"[4]—all on the same power budget as a 20-watt lightbulb.

With the differences drawn between the hardware and software of the aircraft and the *wetware* of the operators, we can begin to prepare the ground for a communications strategy that can be repeated, scaled to fit any size organization, and adapted to fit any style of operation. As you might expect from the masthead quote at the beginning of this chapter, it involves the use of Klein's concept of intent as applied to the automated aircraft systems. There are two simple concepts surrounding communications between aircrew and the aircraft system; they have been known since the earliest research into the operation of highly automated aircraft, but have remained obscure or undeveloped for over two decades. They are the ability of the crew to effectively share information with both the aircraft and one another.

Communicating Intent, Crew to Aircraft

In 1989 we see the first official references (in print) to the now-cliché trio of questions most heard on the automated flight deck: "What's it doing now?" "Why is it doing that?" and "What's it going to do next?"[5] It was then, and remains now, a question of *intent*—the intent of the crew (as expressed in their interaction with the automation) and the inability of the crew to understand the automation as it generates information for them. Make no mistake that in the discussion of "intent" with automated aircraft, the aircraft automation performs along very narrow pathways, and making "intelligent" decisions about what to do next and what the crew might expect is not a typical design criterion. We obscure the automation's true capabilities (and our own) when we attribute these qualities to the flight management functions and autoflight. Calling the automation "magic," "George," or even "it" is a sure sign that the relationship of crew authority over the aircraft is dangerously inverted. In human-machine interactions, the human must remain "in charge" at all times. To fully master these systems and avoid asking the questions above, pilots need to know that the aircraft is either (1) performing along its designed pathway or (2) performing according to a modification entered by the crew. In either case, the results can be desired or undesired, based on what the crew has or has not done to ensure that the automation is behaving according to the *crewmembers'* intentions. This is the role reserved for the wetware—and it must be embraced with discipline and knowledge.

As we have pointed out in previous chapters, the primary interface with the automation controlling the flight path of the aircraft is through the flight guidance control panel (FGCP) and the FMS. Whether using a cursor control device (CCD), multi-function control and display unit (MFCDU), or other input device, the crew has a variety of ways to communicate intent to the automation. ("This is what I want you to do, this is how I want it done, and this is how I expect you to do it.") As air traffic control and airspace management become increasingly complex, more is expected of the automated aircraft systems in maintaining a predetermined flight path in both the lateral and vertical—and increasingly, time—modes. For the automated systems to provide the dutiful service they were designed to, it is imperative that pilots and crew adhere to the following protocol for entries to the FMS, FGCP, and other automation interfaces:

1. Entries are 100 percent errorfree.
2. Entries are cross-checked and verified by the crew against published charts.
3. Entries are cross-checked and verified among crewmembers.
4. FMS and chart databases are checked prior to every flight.
5. Mode engagement is verified when autoflight systems are coupled to FMS modes.
6. Programming of the FMS and other inputs to automated systems are briefed between crewmembers to create a redundant level of supervision over the automation.

Using the FMS over the duration of a flight or mission usually requires constant interaction with the programmed (intended) flight path of the aircraft, laterally, vertically, and possibly with time constraints. Many factors can act independently to derail the protocol outlined above and erode the crew's ability to communicate with the automation over the course of the flight: fatigue, complacency and boredom, emergency and abnormal situations, and a lack of flight discipline can all conspire to interfere with inputs to the automation or can lead to incorrect, inaccurate, or unintended outcomes. The only certain way to counter the constant disruptions to the crew-to-automation communications is to build reliable, repeatable processes into the policies, procedures, and practices (three of the "four P's" addressed in later chapters) on the flight deck. We have already stated that "Setup is everything." To guarantee the best outcomes and realize this attitude, crews must be prepared to act on the messages of "intent" which the aircraft systems dutifully—though sometimes discreetly—communicate to the crew.

Communicating Intent, Aircraft to Crew

We have already witnessed the vain attempts of aircraft systems to capture the attention of their crews and alert them to flight path deviations that would ultimately end in disaster. Air Inter 148's FGCP set to −3300 ft/min on the descent to Strasbourg and Eight-Five Victor Tango's approach mode not capturing during the approach in Houston are just two examples of discreet flight path status "messages" that populate automated cockpits at an alarming rate when compared to traditional, steam-gauge cockpits. Many more accident reports (and reports to numerous safety systems) chronicle the confusion and uncertainty of cockpit crews while the aircraft system dutifully reports not only what *it is doing now*, but also what *it is going to do next*. To emulate the very best glass cockpit pilots among us, we must adopt a mind-set that allows us to take in messages of intent generated by the aircraft, test them against the desired outcome of the crew, and either act on them in confident compliance or initiate immediate intervention to guarantee the safe passage of the aircraft under our control. The way that we categorize these messages of intent will help you to organize the information that streams onto the flight deck of the aircraft you operate, and perhaps provide a step up toward handling the most important information during the most critical phases of flight.

In research dating back as far as 1991, NASA has shown that in the complex interaction between flight crews and automation, the computational assistants in the cockpit also need to communicate intent.[6] Designers continue to apply innovations that more vividly display the intent of automation to pilots; yet, to date there has been no one-size-fits-all solution. We have advocated for years the formation of a disciplined approach to the problem, centered on the crew and not the automation—an approach that leaves the crew clearly in charge. By organizing the "gluttonous banquet of information" crews can make better decisions about what to put on their plate—only the information that is meaningful for any situation

Priority One	
Type of Information Provided by Aircraft Systems	**Examples of Priority One Information**
Immediate, real-time status of the aircraft's flight path and vital major system status (What is it doing now?)	• Lateral and vertical navigation information from EFIS displays • Flight path information displayed on HUD • Basic airspeed, heading, and altitude information • System(s) status • Aural and haptic alerting systems such as enhanced ground proximity warning system (EGPWS), traffic alert and collision avoidance system (TCAS), and stall warning "shakers" • Time and distance information related to vertical and lateral navigation parameters • Advisory, caution and warning system (ACAWS)/engine-indicating and crew-alerting system (EICAS)/electronic centralized aircraft monitoring system (ECAM)/caution, advisory and warning system (CAWS) warnings, cautions, and advisories

Priority Two	
Type of Information Provided by Aircraft Systems	**Examples of Priority Two Information**
Information on future status of the aircraft's flight path and major system status (What is it going to do next?)	• Trend information displayed on PFD, HUD (airspeed, vertical speed, lateral predictions) • Climb, descent, acceleration, and deceleration symbols depicted on the lateral and/or vertical navigation displays • System status messages such as navigation accuracy, navaid status, and degraded aircraft capability • Radar, terrain awareness, predictive wind shear displays • ACAWS/EICAS/ECAM/PFD informational messages on future status

Priority Three	
Type of Information Provided by Aircraft Systems	**Examples of Priority Three Information**
Noncritical flight path planning information, major and minor aircraft system(s) status that does not impact the flight path of the aircraft unless activated or requested by the flight crew (What do we want it to do later?)	• Information on status or synoptic displays (EICAS, ACAWS, etc.) relating to major aircraft systems • Planning information on "secondary," "alternate," "stand-by" or route 2 flight plans • Information on planning pages of the FMS such as fuel, wind predictions, and performance data • ACAWS/EICAS/ECAM informational messages for reference

TABLE 7-1 Aircraft-to-Crew Communications by Priority

on the flight deck and nothing that they do not need, no matter how tempting the dish.

Every manufacturer must comply with certification standards that require their cockpits to categorize, prioritize, and suitably display information generated by the FMS and other mission computers as well as major aircraft system controllers. Additionally, aural and haptic alerting systems (such as voice alerts, "stick shakers," or "stick pushers") are likewise ordered by priority, which is an essential first step in organizing information for the crew. Yet as automatic systems communicate their current status (what's it doing now) and intent (what's it going to do next), the competition for the crew's attention continues as automatic systems coldly and dutifully communicate information to them. Crews must know where to look for the information they need and what it is telling them. In subsequent chapters we examine this issue in greater detail, but for the purposes of communications, crews must first know where to look for what they need and what it means to them immediately. To make this simpler, we have organized all the information generated by the aircraft into one of three levels: priority one, priority two, and priority three communications depending on the importance of the information to the crew's decision making in determining the answers to the questions of what it is doing now, what it is going to do next, and what we want it to do later. (In Chap. 12 we take on the problem of "why it did that.") See Table 7-1.

Each priority level is comprised of information from virtually every system on the aircraft; in other words, the essential information on the flight deck is not prioritized by system or function, but rather by what it means to the crew as they work together with the equipment and one another to make decisions in flight. While this will help in understanding the intent of the aircraft, we must also confront the essential, expert knowledge required by glass cockpit flight crews in communicating with one another.

Crew-to-Crew Communications

Up to this point we have suggested clear and unambiguous processes that crews can use to ready themselves and their aircraft for the demanding, multioperator, high-risk/high-reliability environment modern aviators find themselves in every day. Principles we have shared for briefing and debriefing, data entry (Chaps. 5 and 6, respectively), and aircraft-to-crew communications have created a sound foundation for crews to perform at expert performance levels. We must now complete this equation and update what our industry has long recognized as a key component of expert cockpit performance: how crews interact with one another in the aircraft during routine and nonroutine events which continuously stream into the cockpits and mission compartments of complex aircraft, commonly referred to as crew-to-crew communications.

The research community in recent years has advanced well beyond the basics of understanding the components of communications that our industry has focused on for over two decades: *the words we choose, how we say them, and the nonverbal cues which accompany them.* For our purposes, we trust you have learned this in previous literature or CRM training, for we do not spend time on these basic elements that all experts and most novices already understand. Our goal in these next few pages is to take the preparation skills and expert knowledge of how we set up our crew and the automation to help us execute our goals and forge a high-performing, internally reliable team that functions smoothly and effectively in every task encountered during every flight, under every kind of workload. We are steadfast in the assertion that today's modern flight deck requires greater amounts of a certain kind of communication, reduced amounts of communications which are typical in less sophisticated aircraft, and a repertoire of skills that both pilots (in a two-pilot cockpit) and all aircrew (in multiplace mission compartments) must master and maintain as front-line airmanship skills.

Recent research on communications in expert teams allows us to back up what we have been watching experienced glass cockpit pilots do in this critical skill area for more than 20 years. As an industry we are fortunate to have widely respected human performance scientists working to define both how expert teams are formed and how they perform. In addition, the engineers and designers behind modern cockpit displays and indicators are increasingly leveraging the science of communication to improve the reliability of crewmembers operating advanced aircraft. Professors Christopher Wickens and Justin Hollands exclaim in their textbook on the subject: "The data reemphasizes one theme . . . The design of effective systems for information display and control with the single operator is a necessary but not sufficient condition for effective human performance."[7] Advanced aircraft require crews with advanced knowledge of communications.

In a paper published in 2005, a team of researchers led by Eduardo Salas details the makings of an expert team, titled "The Making of a Dream Team." Not surprisingly, communication plays an important role in shaping the key qualities identified in the research. A list of these characteristics is developed extensively in the research; but those knowledgeable of the demands of the automated flight deck can use the following summary to visualize a highly effective team carrying out these behaviors through clear, concise, and highly coordinated communications between team members.[8]

1. Expert teams hold shared mental models: expert teams are composed of members who anticipate one another's needs.

2. Expert teams optimize resources by learning and adapting: expert teams self-correct, compensate for one another, and reallocate functions as necessary.

3. Expert teams engage in a cycle or discipline of prebriefing → performance → debriefing: expert team members provide feedback to one another.

4. Expert teams have clear roles and responsibilities: expert teams are composed of individuals who manage their expectations by understanding one another's roles and how they work together to accomplish the team goals.

5. Expert teams have a clear, valued, and shared vision: expert teams have a clear and common purpose.

6. Expert teams have strong team leadership: leaders of expert teams are not just technically competent; they possess quality leadership skills.

7. Expert teams develop a strong sense of "collective," trust, teamness, and confidence: members of expert teams are able to manage conflict appropriately by confronting one another effectively.

8. Expert teams manage and optimize performance outcomes: expert teams make better decisions and commit fewer errors. They are able to balance their communication so that team members have the appropriate and timely information they need to contribute to the team, thus creating a higher probability of mission success.

9. Expert teams create mechanisms for cooperation and coordination: expert teams are able to identify all the relevant teamwork and task work requirements and ensure that, through selection and training, the team is composed of individuals possessing the competencies necessary to successfully meet the team and task work requirements.

Many of the functions of expert teams identified by Eduardo Salas and his colleagues have been raised as vitally important functions and detailed in earlier chapters. However, we have to discuss in detail exactly what expert crewmembers say to one another based on this knowledge and how they say it. There seems to be much consensus on the characteristics of expert communication, *but how is it done?* Already we have cited a variety of ways that expert crews communicate prior to a flight or mission, in Chap. 5 on briefing and debriefing and Chap. 6 on data entry. Having laid this groundwork before they are faced with the need to adapt and modify their plans and face problems involving complex problem solving, expert glass cockpit crews are clearly ahead of the game. When things change and problems arise, these crews bring to bear communications skills that are compatible with the technology, airspace, command and control, and cultural environments they are working in. Highly coordinated, effective communication within expert-performing crews is marked by the following, specific kinds of communications:

Expert Glass Cockpit Crews . . .

1. Use standard phraseology during routine and nonroutine phases of flight: the language they use is specific to the automated equipment and its functions. "The ILS and glide slope have captured, preselect the missed approach altitude in the FGCP altitude window."

2. Challenge ambiguity and require clarification for all requests or commands that do not fit with the shared mental model that has been prebriefed and verified by all crewmembers. "We're about to take off in "heading" mode when we briefed a NAV SID. We should select and arm NAV prior to takeoff."

3. Interact with air traffic control and other "agents" with the same level of precision used when addressing fellow crewmembers, using the lexicon of applicable technology in use. "Shanwick Radio, BigJet 1, Position MIMKU, CPDLC, over."

4. Clearly communicate their intentions to one another on how they plan to use automated systems prior to operational maneuvers and instrument flight procedures. "I plan to do a coupled approach to a fully coupled hover to 50 feet prior to deploying the rescue swimmer. I'll remain coupled to the autoflight unless it doesn't respond within the parameters of the maneuver."

5. Ignore traditionally steep authority gradients between crewmembers since they know that the Pilot Flying is primarily responsible for the safety of the aircraft flight path, while the Pilot Monitoring is responsible for effective back-up; crewmembers communicate fearlessly and with mutual respect regardless of their rank or seniority. "I (copilot) plan to hand-fly the departure without autoflight until 10,000 feet unless I feel the workload is too great for you, at which point I'll couple the flight guidance to the autopilot."

These communication skills are hallmarks of expert, glass cockpit crews. We are consistently reminded of one common theme by pilots and crewmembers who successfully complete a transition to a highly automated aircraft for the first time: "I knew that the technology would be different, and there would be a steep learning curve on the automated system . . . but I didn't expect to learn a whole new way of talking and listening with my fellow crewmembers." What these pilots and crews are acknowledging is largely the sense of increased interdependency between each position in the crew: the adaptive, reliable, and highly automated flight deck has allowed for fewer individuals to accomplish the same tasks, but has not had the same effect on the number of tasks themselves. In fact, crews are now more interdependent than ever—in the areas of both task interdependence and workflow. Expert teams not only sense this about their job on the flight deck, but also know it in detail; and once they know it, they begin to manage it, raising their performance level above their colleagues who do not.

A group of researchers in 2008 reported in the *International C2 Journal* on how communications and awareness drive high-performance teams. Their report reads like a laundry list of essential knowledge for glass cockpit flight crews, detailing the kind of communication that these high-performing positions require and how they use this knowledge to achieve high levels of coordination and problem-solving capability.[9] In environments where no one individual has all the required knowledge and skill to accomplish a goal, a team must form and bring together diverse knowledge and skills: "Success in situations such as these involves, in part, connecting and, if possible, uniting people with the power to make decisions and take action with those that have an understanding of the situation and what needs to be done. A decoupling between these two capacities is a source of inefficiency and error."[10] We have all seen this decoupling between what is happening and what needs to be done play out on a dramatic scale when highly visible aviation accidents consume the attention of both those within our industry and those who apply it to their personal and business uses. Yet this decoupling can be seen readily on a smaller scale, even if the errors that ensue do not make the news, result in a safety report, or even are not captured in a postflight debriefing. Expert crewmembers know the relationship and manage this relationship exhaustively as a primary defense against errors and inefficiency. Just how they do so is revealed in the specific, detailed, and deliberate way they communicate their intent with one another and use verbal and nonverbal communications to follow up and act on their plans.

In basic CRM training crews learn the difference between *explicit* (often taught as "verbal" communications) and *implicit* communications (referred to frequently as *nonverbal* communications), and they are encouraged to employ both with care and skill in dealing with one another. This is generally where this training ends—with crews aware of the limitations of spoken words and the power of the unspoken message given off in body language, tone, and even the decision not to say anything at all. We aim to take this training one step further and suggest that it is required of all glass cockpit crewmembers (as we have discussed already) to *speak clearly with the shared language of automation in the context of what the automation has been programmed to do,* consistent with the shared goal of the crew. The overlap between briefing and debriefing and communication should be visible during high-workload, high-coordination phases of flight (such as low-visibility ground operations, precision approaches, critical operational procedures, and emergencies). Less visible to the casual observer, however, will be the smooth and effective way that expert teams interact:

> ... the capacity to shift between explicit and implicit coordination strategies is a hallmark of expert teams; explicit coordination relies on verbal communication...implicit coordination [depends on team members' ability to draw on] shared mental models of the team and task to anticipate the needs of their fellow team members and pass information and other task inputs before they are requested; teams are aware of how to organize their behavior because: 1) they share an understanding of the situation; and, 2) are capable of interpreting this situation in terms of the task needs of their fellow team members...[11]

In this illustration the research community seemingly describes the "unencumbered elegance" of an expert crew performing tasks they have both carefully set up in the automation and briefed in detail. Those of us with experience in glass cockpit aircraft can virtually see this in our mind while reading the characterization of expert performers; those with little or no experience can imagine how they might shape their own communications on the flight deck or mission compartment to match the level described herein.

Already the overlap among the first four Automation Airmanship principles is beginning to demonstrate the interdependencies that crews must balance in advanced aircraft. In Chaps. 8 and 9 we add two critical principles to the four we have already discussed— *monitoring* and *situational and mode awareness*—two natural complements to much of the information we have already presented.

Checklist for Success: Communicating

☑ Brief all inputs, modifications, and changes programmed and/ or input in the automation. *Point* to input or readout locations, and announce automation mode changes to follow up and promote clear communications.

☑ Compare constantly crew intent with the actual performance of the automation.

☑ Understand the difference between information that communicates what it's doing now, what it's going to do next, and what we want it to do later.

☑ Include "intent" when configuring and briefing flight-critical operations and how automation will be used during the maneuver.

☑ All ambiguities and setup uncertainties are confronted and annunciated among crewmembers, without prejudice.

Notes and References

1. Daniel Kahneman and Gary Klein, "Conditions for Intuitive Expertise: A Failure to Disagree," *American Psychologist*, 64(6), September 2009. Washington, DC. p. 523.
2. Gary Klein, *Sources of Power*, MIT Press, Cambridge, MA., 1998, p. 229.
3. Massimiliano Versace and Ben Chandler, "The Brain of a New Machine," *IEEE Spectrum*, 47(12):32, December 2012.
4. Ibid.
5. Earl L. Wiener, NASA Contractor Report 177528: Human Factors of Advanced Technology ("Glass Cockpit") Transport Aircraft. University of Miami, Coral Gables, Florida. Prepared for Ames Research Center CONTRACT NCC2-377, June 1989. National Aeronautics and Space Administration Ames Research Center. Moffett Field, Califomia, 94035.
6. Gary Klein, *Sources of Power*, MIT Press, Cambridge, MA, 1998,. p. 230.
7. Christopher Wickens and Justin Hollands, *Engineering Psychology and Human Performance*, Prentice Hall, Upper Saddle River New Jersey. 2000, p. 234.
8. Eduardo Salas, Michael A. Rosen, C. Shawn Burke, Gerald F. Goodwin, and Stephen M. Fiore: "The Making of a Dream Team, When Expert Teams Do Best" in *The Cambridge Handbook of Expertise and Expert Performance*, K. Anders Ericsson, et al, Eds. Cambridge University Press. Cambridge, UK. 2005, pp. 446–449.
9. Michael A. Rosen, Stephen M. Fiore, Eduardo Salas, Michael Letsky, and NormanWarner, "Tightly Coupling Cognition: Understanding How Communication and Awareness Drive Coordination in Teams," *International C2 Journal*, 2(1), 2008.
10. Ibid., pp. 5–6.
11. Ibid., p. 7.

CHAPTER 8

The Fifth Principle: Monitoring

*Delegation means that we commission other institutions and persons
to do detail work for us but that we remain conscious of the role the
delegated problem has in the overall problem. We stay in touch with the
delegated problem.*

—Dietrich Dörner, 1996[1]

Two significant events involving large commercial transports, occurring 15 years apart, one in Japan and one in the United States, reveal the implications of the failure of flight crews to successfully monitor information on the flight deck provided by their highly reliable automated systems. Although these events were different in many ways, what they tell us about the challenges of effective monitoring on the automated flight deck makes them perfect bookends for one of the most important discussions in modern aviation safety. The first, occurring in 1994, was held up to industry, operators, and flight crews as a harbinger of what might signal an increase in the number of fatal accidents if significant steps were not taken to address the human factors complications that accompany highly automated systems. In fact, following this and other high-profile accidents in the years that followed, leaders in the safety profession would come out emphatically in favor of measures that would give the traditionally marginalized skill of *monitoring* greater emphasis in the training, evaluation, flight deck and procedures design, and the cockpit culture of every operator of complex aircraft. The second event in question occurred long after these changes were implemented, and in full view of the implications of leaving a highly automated system unattended to by its human supervisors. These two examples are not the only events in which poor monitoring would factor into the outcome; they simply are widely recognized and well investigated in comparison to many other events involving a breakdown in monitoring that are routinely reported to safety organizations worldwide. In one case, the outcome would be fatal for the crew and many of the passengers involved; in the second, the outcome would result in speculation and disbelief that the event could happen at all, casting little more than embarrassment and ridicule on those involved (publicly, that is), although the result could certainly have been disastrous.

Failure to Remain Conscious of the Detail Work

In the late afternoon of April 26, 1994, a China Airlines A300-600 operating as China Airlines flight 140 left Taipei, Taiwan, with 271 passengers and crew onboard for Nagoya, Japan. In little more than 2 hours, only seven passengers were alive after the crew, both qualified and experienced pilots, disregarded the information clearly displayed before them and actually acted in opposition to the normally functioning automated systems after the unintended actuation of the aircraft's automatic go-around switches. The first officer's actions as the pilot flying (PF) and the delayed action of the captain as pilot monitoring (PM) created flight conditions that would require a heroic act of airmanship to recover from. In short succession after the actuation of the go-around switches, the aircraft stalled and crashed short of its destination, killing nearly everyone on board. Many aspects of this tragic accident can be used to exemplify much of what we advocate in this book, across nearly every Automation Airmanship principle we introduce. But we will examine it first in the context of the monitoring implications, and how a disciplined, well-informed, and systematic approach to monitoring would likely have led this crew to execute a routine go-around to a second approach to landing, making it less than a footnote in the history of global air operations for that particular day.

Losing Touch with the Delegated Problem

In stark contrast to the Nagoya accident in 1994, a now notorious event in October 2009 involving the crew of Northwest Airlines[2] flight 188 (an Airbus A320 similar to the one described in Chap. 1) has left a permanent stain on the professional reputations of not only the crew involved and the Northwest pilot group, but also the airline pilot profession as a whole. The crew's inattention to their flight deck duties led to the aircraft overflying their destination, while not responding to air traffic control radio calls, company messages sent to their onboard messaging system, and the routine displays on a flight deck designed for ease of monitoring and for notifying the crew of phase-of-flight status. Their actions have been described as a "frolic" and a "total dereliction of duty"[3]—even if the cause was as simple as the breakdown in the crew's monitoring protocols, which are laid out in a multitude of company and professional guidelines on vigilance. Although it was not as exciting as the crew-induced and tragic wild ride of China Airlines flight 140, the passengers of

Northwest flight 188 were, unknown to them for approximately an hour, on a different kind of wild ride—one in which their crew would fail to complete even the most basic monitoring task in aviation, that is, *listening to their radios.*

In both of these events the breakdown in monitoring had unintended results, but the breakdown occurred differently in each case: one during the relatively intense few minutes of an instrument approach at the destination, the other during a comparatively low-workload, en route phase of flight with less immediate information to attend to by the crew. What they have in common, however, and what binds them together in this unique class of accidents, is the role that automation played in its dutiful (and accurate) notifications to the crews of each flight of what was happening—and each crew's failure to properly act on the automation's notifications to them.

Elevating Monitoring to a Front-Line Duty

Before we can understand how these crews may have acted differently if confronted with the same information in a similar situation, we must look into the dynamics of monitoring: what it is, in and of itself, and what we can learn about how experts apply monitoring on the advanced flight deck. Much of what we describe in this book are skills that can be deployed on the front line of traditional airmanship, alongside "stick and rudder" skills that are born of knowledge, practice, and experience. *Monitoring* may be the singular skill that many pilots (and even some experts) possess little knowledge of, and practice with equally little diligence. We cannot inspire every individual to new levels of diligence, but we *can* provide the latest information in order to raise the level of knowledge across the entire industry to that of "expert" master monitor.

Part of the body of knowledge of monitoring that all experts should possess is the view that leading researchers began advocating a decade ago, and that has yet to be fully adopted by industry. In an industry paper presented in 2001, Dr. Randall Mumaw of Boeing, Dr. Nadine B. Sarter of Ohio State University, and Dr. Christopher Wickens of the University of Illinois made the case for developing greater knowledge and proficiency across the industry:

> Given the complexity of the flight deck interface, we believe that effective knowledge-driven monitoring is critical for effective fight operations; even more important than data-driven monitoring. These data suggest that pilots have insufficient knowledge of automation behavior to anticipate important automation state changes. Training to improve pilot mental models and to improve monitoring strategies needs to be developed until flight deck interface improvements can be established in the fleet.[4]

We can begin to build the knowledge we need from what another prominent group of industry experts goes on to advocate at about the same time as the "four underlying causes of poor monitoring."[5]

1. Industry has not identified monitoring as a primary task, but rather has implied that it is a secondary task to other more important flight deck duties.

2. The current system does not "reward" proper monitoring, and vigilance in the interest of safety that can cause delays may be viewed as a nuisance, causing monitoring to have less priority on the flight deck.

3. Monitoring is a skill that appears to be intuitive and easy, when in fact it is un-natural and requires understanding, practice and feedback to be effective.

4. The reliability of systems [and] the infrequency of critical errors can lead to complacency and diminish the effectiveness of monitoring over time.

There are indeed many differences between the way experts and nonexperts perform on the advanced flight deck. One of the most distinguishing qualities that separate the two is the ability of experts to select from the vast amount of information just the right "pieces" of information, at exactly the correct moment, to achieve a seemingly perfect outcome. A trained observer, well versed and knowledgeable of the components of effective monitoring, can readily see the performance of experts' actions that systematically works in opposition to the four primary causes of poor monitoring just identified. These flight crews consider monitoring tasks as primary crew duties, establish and maintain a flight deck climate that promotes and rewards effective monitoring, and consistently apply a deep knowledge of their own ability to monitor across every phase of flight.

In the process, these highly competent flyers seem to move with relative ease through the maze of information, with little distraction, no matter how heavy the workload, and often without heed to the hazards of fatigue and boredom that often accompany long, routine, uneventful flights. Indeed, these experts are able to effectively prioritize information generated by the aircraft systems and in turn communicated to the crew, but they also appear to *think* and *act* differently than their less-expert colleagues in the way they gather and distribute the information that leads them to high levels of performance. Without a doubt, they have developed expert knowledge of the displays, indicators, and warning systems that make up their workspace. But much more than that, they have a deep knowledge and understanding of their own physical and mental capacity to obtain, organize, and act on the information on the flight deck (and often they apply this knowledge to their fellow crewmembers). *It is at this intersection between the information that the technology provides and how experts use it that we find crewmembers who are highly effective at monitoring the technology and managing their own and their crew's attention.*

What "Expert Monitoring" Really Means

For several decades the lexicon for describing how displays and indicators are used on the flight deck has included *delegation, supervision, vigilance, attention,* and *monitoring,* to name a few. Many researchers have conducted numerous insightful studies that give new meaning to these critical crewmember functions on the advanced flight deck. As a consequence, the meaning of monitoring as a skill and discipline has been blurred over the years: as aircrew, we know we *have* to do it, but we do not always know exactly *how* to do it. At their first introduction to a highly automated flight deck, flight crews are often told, with great anticipation, of the workload reduction that the careful design and integration promises, made possible by the crew's ability to, as Professor Dörner suggests, "delegate" tasks to the automated systems, "to do the detail work for us." Typically not much is mentioned about the vital role that monitoring plays in this process (the part about staying in touch with the delegated problem) or even the different types of activities that comprise a successful monitoring strategy. Compared to all the possibilities represented by increasingly elegant design, perhaps the least compelling of all the required training is *monitoring.*

To make this subject not only accessible but compelling, we are going to leverage the source of knowledge that delivers the "essential systems knowledge" for monitoring: the growing understanding of how humans are configured to monitor their surroundings for information that is crucial to their own risk and recovery, and our understanding of the technology (displays, indicators, and alerting systems) that provide notification to the crew of what is happening, what might happen next, and why.

For an individual's and a crew's monitoring activities to work effectively, there must be a common concept of what constitutes this vital skill in the first place. The *Oxford English Dictionary* lists several definitions for the verb *to monitor,* including "to check or regulate technical quality without causing any interruption or disturbance" and "to observe, supervise, or keep under review; to keep under observation; to measure or test at intervals, esp. for the purpose of regulation or control."[6] Synonyms for the word *monitor* include *check, watch, observe, supervise, scrutinize,* and *examine.* In any case, monitoring on the automated flight deck requires that individuals use their physical and cognitive abilities to direct their *attention* toward a specific source of information to accomplish any of those defined tasks. The *attention mechanism*, therefore, is a key component of monitoring. So in turn, we must have a sound and universal model for *attention* before we can expect crewmembers to be effective monitors at all—a much steeper climb than simply defining monitoring with the help of a dictionary, or simply assigning individual crew responsibilities such as pilot flying (PF) and pilot monitoring (PM) and hoping that both crewmembers know exactly what to do.

In the search for a concise model of attention, many distinguished researchers have made significant contributions over the past several decades to achieve a comprehensive yet simple model for attention that can be applied across virtually every human endeavor to make outcomes more effective (everything from browsing titles in a bookstore to operating complex systems). First proposed by Michael Posner, the concept of a *spotlight* to describe attention is a significant leap forward when describing monitoring and enacting it on the flight deck. The spotlight model of attention is explained here by biologist and neuroscientist Francis Crick, co-discoverer of the structure of DNA:

> A common metaphor is that there is a "spotlight" of visual attention. Inside the spotlight the information is processed in some special way. This makes us see the attended object or event more accurately and more quickly and also makes it easier to remember. Outside the "spotlight" the visual information is processed less, or differently, or not at all. The attentional system of the brain moves this hypothetical spotlight rapidly from one place in the visual field to another, just as, on a slower time scale, you move your eyes.[7]

Posner's spotlight model has been adopted across much of the field of cognitive science, and it has been advocated by other leaders in the field since its introduction in the mid-1970s. Twenty years after the spotlight concept of attention was introduced, Dr. Crick suggested that much of what we consider as consciousness is comprised of what is "in our spotlight"[8] and adds to our understanding of the spotlight model: "Because we see more clearly close to our center of gaze, we get more information about an object if we direct our eyes in that direction. We get coarse information (at least about shape) from objects we are not looking at directly."[9] The spotlight model includes both how the brain processes information and how our sensory system perceives (close your eyes and listen to your surroundings—you are still directing your mind's spotlight although you may not be using your eyes). You should, by now, be thinking about how you manage your own spotlight, especially on the flight deck or mission compartment; but there is more. It is important that we distinguish between attention and its close relative, *arousal,* which, in Dr. Crick's words, ". . . is a general condition affecting all of one's behavior, as you may notice when you first wake up in the morning. Attention implies to psychologists, as William James said, 'withdrawal from some things in order to deal effectively with others.'"[10] Simply stated, for purposes of effective monitoring, wakefulness and attention are related only physically.

In the *Handbook of Human-Computer Interaction*, Robert W. Proctor of Purdue University and Kim-Phuong L. Vu of Cal State, Long Beach, refine our understanding of attention, adding to Crick's nod to William James' notion of attention:

> Attention is increased awareness directed at a particular event or action to select it for increased processing. This processing may result in enhanced understanding of the event, improved performance of an action, or better memory for the event. Attention allows us to filter out unnecessary information so that we can focus on a particular aspect that is relevant to our goals.[11]

Again according to Crick, ". . . an *attended* event is reacted to more rapidly, at a lower threshold and more accurately (emphasis added)."[12] By managing your spotlight, you are managing your attention and in turn your very consciousness. If you are doing this with routines on the flight deck, you are indeed making significant steps toward the notion of becoming a "master monitor" whenever the task demands.

But how do experts turn this knowledge of attention into effective monitoring? How can less experienced pilots and crew adopt this approach and reap its benefits in a like manner? As is true with many other habits that experts maintain, they bring their knowledge to the forefront of how they think and act; they know that what is *attended to* will be *done better* than what is not; and when they do everything—from conducting briefings to executing a precision maneuver—they consciously and deliberately direct their own and their crew's spotlight to those things that factor in the safe and effective outcome, before considering lesser priorities (in a similar manner). Previous chapters have identified the best practices of experienced crews in planning, briefing and debriefing, data entry, and communicating, and the relationship between these principles of Automation Airmanship and monitoring is understood by experienced top performers. What is important is not only planned for and thoroughly briefed, but also properly set up in the automation, and any changes are clearly communicated among the crew. All these things make it easier for experienced top performers to decide how to deploy their personal spotlight.

But there's even *more* that the experts know about their own and their crews' spotlight: it works best when it moves on command, and not randomly, as it encounters distraction and interruptions. Indeed, expert monitors know that the physiological aspects of superior monitoring do not lurk in the background, ready to deploy when they are needed; *managing their spotlight is a high-level, overt cognitive strategy meant to give them a performance edge.* They are aware of the sequence of action, even if they lack the detailed knowledge that Posner associates with the cognitive laws surrounding the spotlight's dynamic nature, describing such movement as having three successive processes:[13]

Disengage—Move—Engage.

An expert's knowledge of his or her own attention is one of those subtle but effective skills that contribute to the illusion of "unencumbered elegance" and make it appear that experts indeed "have all the time in the world." They simply are aware of their spotlight and so manage how large or small it is, where it is directed, and for how long. Likewise, they ensure that their fellow crewmembers are deploying their own spotlights effectively as well, complementing their own. Rarely, as we will discuss in later chapters, will experts allow all the available spotlights to be directed at the same location at the same time, thus maintaining a constant, habitual discipline of disengage—move—engage, disengage—move—engage, . . .

Because we are now discussing how experienced top performers think about *how they themselves think and perceive*, we can introduce another skill associated with expert performance: *metacognition.* Metacognition—literally *thinking about thinking*—gives experts a decisive performance edge, especially when it comes to effective monitoring. According to Paul Feltovich, Michael Prietula, and K. Anders Ericsson, metacognition is an important aspect of expert performance because ". . . this kind of monitoring prevents blind alleys, errors, and the need for extensive back-up and retraction, thus ensuring overall progress to a goal."[14] Experts have learned, mostly from experience, that there are limits to attention, and these limits must then translate into effective monitoring to prevent them from encroaching on their own performance and, in turn, their crew's performance. They *monitor their own monitoring* (as crazy as it sounds, that is just what they do). Expert glass cockpit pilots are aware of their attentional limits. These limits include

1. That individuals can only focus the spotlight on one problem (or making one decision) at a time when performing an unfamiliar task.

2. To share attentional resources, individuals must shift rapidly from thinking about one task to another task.

3. There are perceptual limits to what the "eye-brain" can perceive, and what we can hear, making it possible (and often likely) that we can miss something in our spotlight.

4. We act and react in a constrained perceptual environment, forcing us to constantly balance attention and awareness.[15]

It is a short leap from our knowledge of the spotlight to effective interaction with the main components of the automated flight deck. Experts show us how, by configuring the information systems to display only what is necessary, within the sensory bands that they can effectively detect (for example, brightness, range, and volume), and demanding the same discipline from their fellow crewmembers. In this way they manage the foreground by excluding nonessential information, limiting the spotlight to only the flight-critical information that matters for the phase of flight that the aircraft is in. Here are a few examples:

1. An expert captain or aircraft commander sets her weather radar range, tilt, and scale to optimally display threatening weather for takeoff, while directing the copilot to set his corresponding EFIS display to "terrain" mode to display dangerous terrain information generated by the EGPWS system along the departure track on departure; she briefs this setup with the copilot and thus calibrates the respective spotlights of both crewmembers for departure.

2. The pilot flying (PF) on a low-visibility approach "declutters" her HUD of all but essential flight path information, making her spotlight reflect the most important information for the approach phase of flight.

3. A relief pilot called to the flight deck during a long over-water cruise segment, upon relieving the copilot, checks the setting of his radio volumes, status of data link, and long-range settings of the weather radar and range scale of the navigation display of the EFIS in order to put all vital information is his spotlight.

4. During an approach briefing, a crewmember briefs the performance criteria of the autoflight system and other critical flight parameters as well as the requirements of the pilot monitoring (PM) to specifically monitor and identify deviations to glide path, final-approach course, sink rates, airspeed, and thrust settings that would require immediate correction or go-around by the pilot flying (PF).

The display technology that is available on today's aircraft simply requires crews to know more about their own cognitive processes than ever before. (We believe that as the importance of effective monitoring has grown while the number of crewmembers on the flight deck and mission compartment has shrunk, more research would have an important impact on all high-risk/high-reliability domains.) Professor

Christopher Wickens, an expert on human-computer interaction (commonly referred to as HCI) and aviation display technology has stated that it is ". . . clear that great improvements are possible by combining theories of cognitive psychology with increasingly available computer and display technology."[16] Keeping up with the advancing technology and the science of HCI would tax most front-line aviators on an impractical scale; however, maintaining a discipline of awareness of one's own monitoring skills and practicing them at every opportunity when involved in specific monitoring functions are peculiar behaviors of experts, which can be adapted to any modern cockpit and its family of displays, indicators, and warning systems (in fact, your next drive home from work will be safer if you apply the same strategies).

It is not our goal to encourage or portray an image of the highest-performing crewmembers being in a constant state of vigilance, ready to spring into action at the first indication of deviation in a complex, highly defended system such as flight operations. Because of physical and cognitive limitations of flight crews, it is necessary to adjust one's attentional system to the situation at hand, for example, high levels of vigilance during ground and taxi operations at a busy airport or during takeoff and landing. It is clearly not practical to maintain the high level of monitoring that this situation most always demands over the duration of even a short mission, let alone a long intercontinental flight. But experts *never* stop monitoring—they simply adjust what critical flight information must be monitored and the rate at which they move their spotlight from information source to information source. Professor Wickens and colleagues have a model for this as well that experts employ across all their operations. We have adapted this model for applications on the advanced flight deck, as experts commonly do.

The workload requirements of a typical flight or mission can be compared to a *conveyor belt* on an assembly line that changes speed throughout a worker's shift. At a high rate of speed (during takeoff and initial climb, for example) the spotlight is narrowed to the most critical information, and automatic (well trained) responses and strategies are relied on to contend with abnormalities. As the belt slows (during climb and subsequently at cruise, say), the spotlight is deliberately broadened to encompass additional sources of information (speed and altitude constraints associated with upcoming waypoints, and air traffic control's reduced vertical separation minimum [RVSM] constraints, for instance) that do not require immediate response, allowing more elaborate decisions to be considered or even discussed among crewmembers. At its slowest rate, the belt slows to a sustaining speed (similar to long-range cruise, holding, extended vectors), allowing even greater consideration of options and likely—and even *unlikely*—events or outcomes. Here is all that the experts know they must do when it comes to monitoring: keep their own and their fellow crewmembers' spotlights on the information that matters *right now*, and ensure that they do not switch off or disengage until the airplane is safe at its destination with the parking brake set, adjusting their spotlight to match the rate of change across the entire flight or mission.

It is important to understand that there is *only one* attentional system; it functions along very narrow, specific lines; and each individual aviator controls his or her own spotlight. Experts do not rely on one type of vigilance during high-workload phases of flight, and then deploy a separate system for the less physically and cognitively demanding phases of flight. Aviators, or other professionals in high-risk/high-reliability professions, do not shift from "active" to "passive" monitoring states; the only thing that changes is the situation—experts simply match their attentional systems to the environment, control what factors they have direct command over, and do not allow themselves to be overtaken by the monitoring demands present at any given time.

We would be irresponsible and unrealistic to suggest that monitoring over the course of a flight or mission is not difficult to sustain in the face of fatigue, boredom, and complacency. Experts know this, too, about monitoring. Professor Wickens goes on to say this about the workload-reducing nature of much of the automation found on advanced aircraft: ". . . this workload-reducing feature can invite problems with automation." That problem is complacency, and complacency can sometimes be fatal. Wickens goes on:

> . . . complacency does not create a problem until automation fails, and such a failure, although often unlikely, is never impossible. One of the ironies of automation is that the more reliable it is, the more it is trusted, and the more complacent the operator becomes.[17]

Because experts "think about their own thinking," they know that, as Wickens continues, "complacency . . . is a cognitive state." Complacency has two consequences for pilots and aircrew.[18] First, failures become harder to detect when crewmembers are complacent. Second, as automation does its job reliably over time (part of the reason why the conveyor belt "slows down" so much on the advanced flightdeck), pilots and aircrew tend to monitor less effectively, and lose situational and mode awareness of the evolving state of the automated system. As we will discuss in Chap. 9, this creates a situation such that when the failure does occur and is detected, the crew will be less able to respond than if they were on watch for the failure in the first place.

We have observed experienced crews using a variety of techniques to stay engaged in monitoring during low-workload phases of flight; setting a cockpit timer to sound a backup aural alarm, using certain functions of the FMS to add visible reminders of necessary cockpit tasks along the magenta line, and deliberately breaking conversations to "page through" systems displays on the EICAS, and check fuel burn and winds are just a few examples. One of the most effective methods we have observed is that experienced top performers *brief* monitoring requirements and expectations prior to long flights or missions with anticipated long durations of low-activity demands for the crew.

To demonstrate both extremes of monitoring requirements on the flight deck, we can now return to the opening narrative of this chapter, that of China Airlines flight 140 at Nagoya and Northwest flight 188 at Minneapolis. But this time, we look at these occurrences through the lens of expert knowledge of effective monitoring: managing one's attentional system via the attention spotlight and a sound understanding of the other portion of the monitoring model we have suggested, the conveyor belt.

When Things Change Quickly

The breakdown in effective monitoring for the crew of China Airlines flight 140 can be traced to the inadvertent actuation of the GO lever by the first officer (FO), who was the pilot flying during the approach to Nagoya. This action caused the aircraft to change flight guidance modes, from approach to go-around, which would have, under normal circumstances, called for the addition of thrust to go-around levels and a change in pitch toward the missed approach altitude, some several thousand feet above the airport. We do not know the reason for the inadvertent actuation, but from the actions that immediately followed it is safe to conclude, along with the accident investigation team, that the FO did not follow through with the appropriate go-around procedure, leaving us with only one conclusion, that actuating the GO lever was an error in the first place.

As the PF during an approach, the FO's spotlight, we must assume, was trained on those flight-critical cues such as airspeed, descent rate, heading and course, and the flight guidance information (now in go-around mode after the accidental actuation of the GO lever). As the pilot monitoring during the approach, the captain would have cast a broader attentional spotlight, which would have included the critical flight path information, but also outside cues such as the runway environment, landing clearance, other traffic in the vicinity, major aircraft systems information, radio calls, and other peripheral sources of data. The actions that would follow would leave investigators convinced that this crew demonstrated, among other things, a lack of understanding of the automated systems of the A300 in this particular situation. Whatever lack of understanding of the aircraft automatic systems this crew was guilty of, their gravest mistake was focusing their spotlight away from critical flight path information that would have demanded they do *something* other than what happened next. That *something* would have been to execute a normal go-around, abandoning their approach, or reconfiguring the flight guidance and autoflight systems from the go-around guidance that was being generated by the aircraft. This would have had the added benefit of "slowing down the conveyor belt" by focusing the attentional resources of the crew to what is normally a well-rehearsed, procedural maneuver. Even for pilots with a deep understanding of their aircraft's automation and exceptionally proficient in the physical steps to configure the FMS, FGCP, and MCP, the latter would have been a challenging, one-chance-to-get-it-right effort. Neither of these options was pursued by the crew. Clearly, the captain was not expecting a go-around, much less the unintended actuation of the GO lever (impacting his situational and mode awareness, a principle discussed in Chap. 9). In the time it took the first officer to accidently actuate the GO lever, the "conveyor belt" of attention had sped up dramatically for both pilots. Their spotlights were on "land!" while the aircraft flight guidance and autoflight systems were set on go-around.

But what happened next defies what expert monitors expect to do when the pace of flight deck activities suddenly picks up: their spotlights remained fixed on the cues that would lead to a normal landing, in spite of what the information was displaying before them: the captain imploring the FO to stay on glide path to landing, and the FO dutifully pushing the control wheel forward against strong resistive force provided by the autopilot, which for some reason had been engaged by either the captain or FO (perhaps as part of a strategy intended to slow down the conveyor belt). The FO's actions on the control column would produce an out-of-trim condition in which the A300's elevator was trimmed to a full nose-up setting—exactly as it was designed to do in such a situation—which would have also been apparent to the crew through both the control force required to keep the aircraft coming down the descent path toward the runway and the information on the instrument panel and aural tones that would have shown the developing abnormal trim situation. In the face of information clearly depicted in front of them, the spotlight of the captain was firmly on landing, that is, until the situation progressed beyond the ability of the FO to control the aircraft, leaving the captain with only one choice: to take control of a large transport that was not only close to the ground, but also badly out of trim. With the conveyor belt now at high speed, and the aircraft pitching to over 40° nose-up,[19] the captain was unable to control the flight path of the airplane which resulted from the out-of-trim condition, leading to the subsequent stall and impact with the ground.

This crew's basic monitoring skills (along with some other issues raised by investigators) during a normally high-workload phase of flight broke down catastrophically, leaving many questions for investigators, families of the victims, and the industry at large. In fact, a full volume of instruction for glass cockpit crews could be derived from the details of this accident, although we only view it here through the specific lens of monitoring (we will leave it up to readers to apply their knowledge of the remaining eight Automation Airmanship principles to this instance). In this narrow context, however, we can appreciate the strategic importance of having a clear understanding of monitoring requirements, by respective role on the flight deck, during the especially important and often-dynamic approach phase of flight.

What happens, then, when the pace is much less intense and borders on repetitive monotony for the crew?

The Monotony of Highly Reliable Systems

At the beginning of this chapter we said that these two events—China Airlines flight 140 and Northwest Airlines flight 188—were ideal bookends for the spectrum of monitoring required by flight crews. Not only can most of us recall an instance during our training when our instructor set the conveyor belt at high speed (and many of us can think of real-life situations in which the same is true), but also those who have logged enough hours to become certificated instrument pilots can recall their fair share of long, uneventful, monotonous flights when they struggled just to be able to monitor the radios. The challenge for many of us, whether flying solo or as part of a large integrated mission crew, is to adhere to strict monitoring protocols when everything around us seems to signal "there's nothing happening now and nothing is likely to happen in the future, so relax." The truth is, you *can* relax, but you have to know how much. Crews can reduce monitoring vigilance during low-workload phases of flight proportionate to the high state of readiness maintained during departure, when over the target, during an abnormal or emergency situation, or during a low-visibility approach; yet in reducing readiness by several degrees, expert crews stay "in touch" with the delegated tasks (recall the techniques we just mentioned). What crews *cannot* allow to happen is to redirect or "turn off" their spotlight completely or to witness the same (or even worse, enable it) in another member of the crew.

But that is exactly what the crew of Northwest flight 188 did on October 27, 2009, on a routine flight from San Diego, California, to Minneapolis, Minnesota. Twenty-five minutes after level off at a cruise altitude of 37,000 ft, the crew of flight 188 would make its last radio call for another 77 minutes while being distracted from their primary flight duties. What the distraction was remains unclear, although official interviews of the crew describe the primary distraction as a discussion about pilot scheduling, including the use of a laptop computer that obscured the crew's view of the instruments.

To miss the multitude of radio calls from air traffic control, numerous attempts by other aircraft and Northwest Airlines Dispatch via the Aircraft Communications Addressing and Reporting System (ACARS),[20] and the variety of alerts that would have been generated by the FMS and displayed as cues on the magenta line, alerting the crew to their flight's "top of descent point" based on in-flight conditions monitored by the aircraft's onboard systems, is clearly the height of complacency.[21] It's hard to imagine even a novice crew so removed from their crew duties by routine distractions that they would intentionally block the view of their own instruments. An expert crew, when confronted with low-workload, low-demand phases of flight, automatically recalibrates their spotlight(s), sometimes by briefing, but most often by procedures established in standard operating procedures (SOPs), or by the aircraft flight manual itself. To expect this level of vigilance is not unrealistic, even in the face of the proliferation of (seemingly) autonomous, self-monitoring, and often self-correcting systems from the autopilot and autothrottles to the major aircraft system controllers designed to alert crews when the aircraft is in or approaching a dangerous condition.

Monitoring and the Maintenance of Basic Airmanship Skills

For generations pilots have faulted autoflight systems as the primary factor in the atrophy of basic airmanship skills—those "stick and rudder" qualities of every pilot's abilities that have long been the source of personal pride and distinction among aviators. We will not argue with the premise that any crewmember who fails to "stay in touch" with the delegated problem will most likely suffer a loss of skill over time. We do elect to differ, however, in assigning blame for this form of complacency to overautomated flight decks and the emphasis in training on using automation to control the aircraft in most phases of flight. In fact, we believe that one of the most vital aspects of any pilot's personal regimen of practice and proficiency is the maintenance of highly effective monitoring skills.

Basic piloting skills rely upon a proper and reliable scan of aircraft status information, particularly airspeed, heading, altitude, and attitude. The basic *T scan* developed from within the industry in the middle of the twentieth century describes the location of the four primary instruments for controlling an aircraft's flight path: across the top of the T from left to right are the airspeed indicator, the attitude indicator, and the altimeter, while the vertical portion of the T consists of, from top to bottom, the attitude indicator and the heading indicator. Since the 1950s this has been the basic layout of these key flight instruments, and this arrangement has been retained by most glass cockpit EFIS displays that dominate today's advanced flight decks. A well-practiced scan of these instruments is at the center of what allows a pilot to be proficient in his or her piloting skills.[22] When automation is engaged, it is essential that pilots maintain this *same scan*, according to the protocols outlined in this chapter. If this basic airmanship scan is routinely maintained, *even when the automation is engaged*, the pilot will find that when the automation is subsequently disengaged (intentionally or unintentionally due to a malfunction or failure), he or she is readily able to return to "hand flying" using the basic piloting skills required. We elaborate on this concept in greater depth in Chap. 11, specifically in how expert pilots smoothly respond to the disengagement of autoflight systems, whether this occurs unintentionally or is a planned action of the crew.

The final part of this section of the book deals in part with a skill that has been called one of the "capstone outcomes" of the Historical Airmanship Model® and has received a substantial amount of attention as well as research across many high-risk, high-reliability industries. *Situational awareness* has been said to be the key to successful outcomes in aviation for decades, yet it cannot be developed, maintained, and constantly updated without highly effective monitoring skills such as those we have outlined in this chapter. We have shown that no matter what the pace of activities on the flight deck, experts never let their own monitoring mechanism rest. The "spotlight" model of attention is not simply a convenient way to describe an essential flight deck behavior; it is fundamental to how human beings have adapted to make sense of their surroundings. By integrating this model with the highly complex, highly automated workplace represented by the modern flight deck, we hope to raise the prominence of monitoring as a vital, well-defined front-line skill.

Checklist for Success: Monitoring

☑ *Monitor the monitoring.* Be aware of the attentional spotlight and the speed of the "conveyor belt." Adapt both to optimize the monitoring of flight-critical information appropriate to the phase of flight

☑ Include critical monitoring roles and expectations in briefings prior to high-workload flight deck activities. Prior to long periods of relative inactivity, brief expected monitoring requirements of each crewmember.

☑ Know and stay within cognitive limits; do not buy into the multitasking myth—consciously maintain a pattern of disengage, move, engage when involved in cockpit activities that require the monitoring of multiple sources of information.

☑ Practice flight-critical procedures, thus freeing cognitive resources for other monitoring chores during high-demand maneuvers or phases of flight.

☑ Devise meaningful routines and rituals for extended cruise segments that keep the "conveyor belt" at a speed that is suitable for keeping complacency off the flight deck.

☑ Communicate with other crewmembers when you are unable to manage your monitoring tasks.

Notes and References

1. Dietrich Dörner, *The Logic of Failure: Recognizing and Avoiding Error in Complex Situations*, Basic Books, New York, NY, 1996, p. 56.
2. In October 2009, Northwest Airlines was in merger proceedings with Delta Airlines and has since been absorbed into that airline. In 2009 the flight in question was operated by Northwest Airlines crew as Northwest Airlines flight 188.
3. FAA letter dated Oct. 27, 2009, from the Southern Region of the Office of the Regional Counsel, to the Captain of NWA flight 188, October 21, 2009, Emergency Order of Revocation, p. 4.
4. Randall Mumaw, Nadine B. Sarter, and Christopher Wickens, Analysis of Pilots' Monitoring and Performance on an Automated Flight Deck, Presented at the 11th International Symposium on Aviation Psychology. Columbus, OH, The Ohio State University, 2001, p. 6.
5. Captain Robert L. Sumwalt, III (Chairman, Human Factors and Training Group, ALPA), Captain Ronald J. Thomas (Supervisor, Flight Training and Standards, US Airways), and Key Dismukes (Chief Scientist for Aerospace Human Factors, NASA Ames Research Center), "Enhancing Flight-Crew Monitoring Skills Can Increase Flight Safety," 55th International Air Safety Seminar, Flight Safety Foundation, November 4–7, 2002, Dublin, Ireland.
6. *Oxford English Dictionary.*
7. Francis H. C. Crick, *The Astonishing Hypothesis: The Scientific Search for the Soul*, Simon and Schuster, New York, 1994, p. 62.
8. Ibid., pp. 23–33.

9. Ibid., p. 59.

10. Ibid.

11. Robert W. Proctor and Kim-Phong L.Vu, "Human Information Processing: An Overview for Human-Computer Interaction," in Andrew Sears and Julie A. Jacko (eds.), *The Human Computer Interaction Handbook: Fundamentals, Evolving Technologies and Emerging Applications,* Lawrence Erlbaum Associates, New York, 2008, p. 54.

12. Francis H. C. Crick, *The Astonishing Hypothesis: The Scientific Search for the Soul,* Touchstone, New York, 1994. p. 59.

13. M. I. Posner and D. E. Presti, "Selective Attention and Cognitive Control," *Trends in Neuroscience,* 10: 13–17, 1987.

14. Paul J. Feltovich, Michael J. Prietula, and K. Anders Ericsson., "Studies of Expertise from Psychological Perspectives," *The Cambridge Handbook of Expertise and Expert Performance,* K. Anders Ericsson et al. (eds.), Cambridge University Press. Cambridge, England, 2007, p. 56.

15. Ibid., p. 57.

16. Christopher D. Wickens and Justin G. Hollands, *Engineering Psychology and Human Performance,* 3d ed., Prentice Hall, Upper Saddle River, N.J., 2000, p. 538.

17. Ibid., p. 544.

18. Ibid.

19. Aircraft Accident Investigation Report, China Airlines Airbus Industrie A300B4-622R, B1816 Nagoya Airport April 26, 1994. Aircraft Accident Investigation Commission, Ministry of Transport. July, 1996, pp. 3–16.

20. ACARS is a radio-based system installed in most commercial air carriers that allows real-time messaging and other operational information to be accessed and displayed through an onboard system in the cockpit, managed by the flight crew. Its use is so common during operations that it is considered routine to send and receive numerous messages during the course of a single flight, and normally it alerts the crew to a message via chimes and flashing alerts on the flight deck. In the case of NWA flight 188, the ACARS messages would most likely have been accompanied by a bell-like chime and a flashing alert on the aircraft's EFIS.

21. NTSB narrative of the incident, DCA10IA001, obtained at www.ntsb.gov, pp. 1–9.

22. A variety of instrument scans for instrument flight are clearly outlined in FAA-AH-8083-15A, *Instrument Flying Handbook,* 2008, Chap. 4.

CHAPTER 9

The Sixth Principle: Situational and Mode Awareness

Since the experts have a mental model of the task, they know how the subtasks fit together and can adapt the way they perform individual subtasks to blend in with the others. This makes their performance so smooth they do not even feel that they are performing subtasks because the integration is so strong.

—Gary Klein, 1998[1]

For over two decades, the concept that pilots and crews can widen their safety margins simply by increasing their awareness of flight conditions has enjoyed broad endorsement by both flyers and those in support roles on the ground. *Situational awareness*[2] (SA) has become such a universally applied term in aviation (and in many other high-stakes activities as well) that it is now widely understood to be a "must have" component of all high-risk/high-reliability endeavors. Invariably, when crewmembers experience a successful outcome following some malfunction or emergency, they are judged to have demonstrated "high SA." Likewise, when they fail to achieve such ends, it is said that their "lack of SA" greatly influenced the undesirable outcome. It is surprisingly ironic to many engineers and managers that the very systems which have been so carefully designed, flight-tested, and rigorously explained to users *specifically to improve situational awareness* have played such a significant role in major aviation accidents. The stubborn persistence of the loss of SA in the cold accident testimony of digital cockpit voice and flight data recorders continues to generate research and dialogue across the industry. Yet the loss of SA persists, even as increasingly durable and intuitive systems designed to improve and increase it are approved and installed every day. Consequently, the human component of the human-machine interface has been a frequent target of research into SA, where it is more difficult to isolate and eliminate failure, or even more difficult to isolate and replicate the mechanisms that build SA within the inherently human dynamic of flying complex aircraft. From all the evidence, it seems infinitely easier to mislead the human component of the system while the solid-state components dutifully hum along as designed, building the case for even greater controls and limits to constrain the wetware within the system. A complete and standardized prevention strategy for the loss of SA still eludes researchers, designers, and engineers. This is exactly what has led us to ask, What do the experts know about maintaining a high level of SA, and how do they apply this knowledge on the highly automated flight deck? Simply, it is more than SA that they are working tirelessly to build and maintain.

What Is a "Mode," Anyway?

In Chap. 2 we described the role of the major components of the automated flight deck and mapped them for easy reference in building a sound understanding of how automation is integrated within an advanced aircraft. Throughout the first eight chapters we have relied on the reader's knowledge of the different "modes" that the autoflight operates in, almost universally, across the entire family of advanced aircraft, with some special exceptions. It is now time to take the path of the experienced top performers and enhance the understanding of exactly what a mode is and how this understanding leads expert glass cockpit pilots toward increased performance.

A "mode . . . [is] a mutually exclusive set of system behaviors,"[3] according to one research group, while Victor Riley, an automation researcher and human-centered designer for Boeing Commercial Aircraft Group, defines a mode as "[a collection of] system actions associated with different system states."[4] Although these definitions are inherently useful to systems engineers, researchers, and FMS design teams, they are not (nor were they intended to be) very practical explanations for pilots and aircrew. However, we can bridge this knowledge gap by applying some terms that are already quite familiar to pilots. If we center our explanation of Riley's definition of a mode on the aircraft's flight guidance control panel (FGCP), we can more easily associate the concept of mode with "commands" that pilots issue to the flight guidance system for *speed, roll,* and *pitch.* If we now add the word *mode* to each of these three primary means of controlling the aircraft's flight path, the result is the three general terms that describe system actions along the aircraft flight path:

Speed mode
Roll mode
Pitch mode

Each of these modes, in turn, has a specified set of behaviors associated with it. To illustrate the practicality of how a mode enables the crew to manage the aircraft flight path, we can look at some of the possibilities related to just aircraft pitch. The pitch mode can be set to hold a specific pitch angle, vertical speed, and angle of climb or descent; and the pitch mode can be guided by vertical navigation systems both internal and external to the aircraft (as in the ILS glide slope or FMS-generated descent path)—all described as the respective "pitch mode" that has been commanded by the crew. These are all specific *system actions* for the *system states* of aircraft pitch. The same approach can be taken for roll and speed modes, according to the functions allowed by the respective aircraft FGCP. A list of flight guidance modes that can be found on most commercial aircraft most often includes those shown in Table 9-1.

How any particular flight guidance mode acts is described in an important research paper presented to industry in 2003 by Anjali Joshi et al.; this research team stated that a mode which is associated with control of the aircraft flight path may act accordingly:

> A mode is said to be selected if it has been manually requested by the flight crew or if it has been automatically requested by a subsystem such as the FMS. The simplest modes have only two states, "cleared" and "selected." Some modes can be armed to become active when a criterion is met. In such modes, the two states "armed" and "active" are sub-states of the "selected" state.[5]

Extending the example of setting aircraft pitch to the definition provided above, the vertical speed pitch mode can be set (on the FGCP) to, say, 500 feet per minute to "capture" the arrival descent path calculated by the FMS, which the aircraft will then fly until it in sequence captures the altitude set by the crew in the FGCP altitude

Autothrottle, Autothrust, or Speed Flight Mode Annunciator (FMA)	Lateral or Roll FMA	Vertical or Pitch FMA
THR or Thrust	HDG or Heading	PROF or VNAV
Hold	TRK or Track	Hold or ALT
Idle or Idle Thrust or THR Idle	NAV or LNAV	V/S
SPD or Speed or Mach	LOC	CLB
Pitch	VOR	DES
Low-speed protection	Takeoff	FPA
High-speed protection	Align	G/S
Retard	Rollout	TO/GA

TABLE 9-1   Common Flight Path and Flight Guidance Modes for Transport Aircraft

window, whereupon it will level off and, if selected by the crew, maintain an FMS-generated descent speed. Of course, all this is done in accordance with the unique design of the particular aircraft, but it is easy to see how the crew is actively engaged in choosing various modes to control the aircraft flight path. There are too many combinations of flight modes that are used for the wide variety of conditions encountered in flight to describe here, but you get the idea.

Now that we have covered the basics of modes associated with control of the aircraft's flight path, it is easy to make the leap to other automated systems on the aircraft, to include flight controls, aircraft powerplants, and installed aircraft systems such as TCAS (traffic alert and collision avoidance system), EGPWS (Enhanced Ground Proximity Warning Systems), PWS (Predictive Wind shear), WAGS (Wind shear Alerting and Guidance System), TAWS (Terrain Awareness and Warning System), RAAS (Runway Awareness and Alerting System), and other advanced systems *designed to increase the situational awareness of the crew.*

As you might expect, modes of all kinds continue to proliferate across the entire family of production aircraft, globally, adding to the information that crews are required to master. Just a few comparisons among popular models of advanced aircraft in wide service today find that the McDonnell Douglas MD-11 provides approximately 67 modes of all kinds, Boeing 777 provides approximately 34 modes of all kinds, and the Airbus A330/340 provides approximately 61. These numbers only are meant to imply that the number of modes available to the flight deck crews of advanced aircraft can vary widely across manufacturers. As a crewmember, it is not necessary to understand the design specifications of every mode that is available to assist in controlling the aircraft or assessing the situation surrounding the aircraft flight path. But *it is essential to understand what a mode is, what a mode does, and what modes are available to the crew during every phase of flight.* The best-performing experts do, and their flight deck actions reveal this knowledge in the smooth, safe, and efficient outcomes that they are so admired for. Among the highest-performing, expert crews, the knowledge and awareness of which modes are active, armed, or preselected is so tightly woven into their "... perception of the elements in the environment within a volume of time and space, the comprehension of their meaning, and the projection of their status in the near future..."[6] that SA and MA have, effectively, merged into one.

Mode Awareness

For the past two decades the uniqueness of mode awareness has attracted the attention of researchers and engineers, yet it has not been promoted as widely as SA in training or evaluation programs across the industry. Occasionally, a notorious accident or near-accident in which a loss of mode awareness played a prominent role will receive some attention by professional associations, training programs, or safety organizations (and their media arms), but those are soon overtaken by other attention-grabbing events in the industry. And SA is too broad a category to describe the role that various aircraft modes play in the outcomes of many contemporary aircraft accidents, incidents, and flight path deviations. It is long overdue for the industry to create a more detailed understanding of how the automation, through the employment of various modes, acts to increase or decrease the overall SA of the crew. To this end, we are reintroducing a decade-old, universal definition of mode awareness so that we can demonstrate the close relationship between situational awareness and mode awareness. Mike O'Leary of the British Airways Safety Service has described a sound, general definition of *mode awareness* that can apply categorically across all aircraft types: "the accurate assessment of aircraft configuration, flight and powerplant parameters, flight control system modes, and the dynamic aspects of all of these."[7]

To achieve the level of integration that high-performing crews already demonstrate, we propose that all pilots and crews merge SA and MA into one collective term, *situational and mode awareness* (SMA), and allow themselves to become more integrated with the overall system which every regulatory agent in aviation has appointed them to master.

Situational and Mode Awareness

The Commercial Aviation Safety Team, also known as CAST, is an industry consortium formed in 1998 with the specific goal to reduce the commercial aviation fatality rate worldwide. Using strict data collection methods, CAST in 2008 reported to industry a host of strategies that could be immediately adopted by airlines and flight operations around the world to improve fleet safety. Making recommendations targeted directly at improving SA across all organizations, both large and small, they stated, "Policies should include statements about the importance of maintaining situation awareness and, particularly, mode and energy awareness."[8] CAST spent years researching SA, MA, and *energy awareness* (EA) and the impact of each on accidents. Their report is one of the most concise references available for rapid and immediate change for organizations facing the prospect of rapid adoption of advanced aircraft. They have weighed in also on mode awareness and energy awareness with the following bold challenge for flight crews:

> Situation awareness requires that pilots know the available guidance at all times. The FCU/MCP and the FMS CDU are the primary interfaces for pilots to set targets and arm or engage modes. Any action on the FCU/MCP or on the FMS keyboard

and line-select keys should be confirmed by cross-checking the corresponding annunciation or data on the PFD and/or ND (and on the FMS CDU). At all times, the PF and PNF [Pilot Not Flying] should be aware of the status of the guidance modes being armed or engaged and of any mode changes throughout mode transitions and reversions.[9]

Their guidance is superb, and much of it is, or has been, adopted by the most professional of flying operations. Yet, we still believe that there is a shortcut to getting crews to look not individually at each concept, but collectively to create the tight integration of these concepts displayed by the most expert of flight deck crews. So, borrowing in part from decades of research as well as observations and first-hand work in the human performance industry, we suggest this fundamental description of SMA:

> The accurate, useful mental model of relevant aircraft automated tasks, including configuration, flight and powerplant states, flight guidance, flight control and sensor modes, and their dynamic relationship to the present and future flight path of the aircraft.

The clear connections to the legacy definitions of SA and MA that have guided experts for nearly two decades are evident and necessarily integrated in this description of SMA. Although the accidents and near-accidents examined to this point have each been used to emphasize the role of other Automation Airmanship principles, it is easy to view each through the lens of SMA to be able to extract valuable examples of the role that SMA played in the outcome. Air Inter 148, China Air 140, 85VT, and US Air 1549 could each provide compelling insights in the context of SMA. (Read the definition of SMA above, in the context of each of these accidents, and you will see for yourself how practical this concept can be in evaluating the performance of both expert and nonexpert crews.)

The current state of the art in expert performance on the glass cockpit flight deck owes much of its content to past research, accident investigations, and the many thousands who have made it their life's work to advance safety in aviation, on the ground and in the air. In the research field for the past two decades, much of the work surrounding SA and MA has fed the scientific community which in turn has informed the manufacturers of equipment, and comparatively little has been provided to the pilots and crews who operate the actual aircraft. Professional pilots do not attend human factors symposia in large numbers, but they do (at an attendance rate of near 100 percent) participate in required annual training, which is intended, in part, to provide them with tools and techniques to improve their individual and team performance. This simple next step is not a great leap, but rather a simplification of concepts that have been converging for many years, and one that brings all crewmembers closer to the objectives that guide their colleagues in the research and scientific communities. We will now show the vital role of SMA as it plays out in the high-stakes, high-drama real world of advanced technology flight operations, and how every aviator can move this into action immediately, alongside other measures deployed routinely in the fight against error that increasingly assails the safety margins of all flight operations.

Experts, Mental Models, and SMA

In Chap. 4 we briefly discussed SA, planning, and the importance of well-developed SA among highly trained and experienced pilots (specifically, the check airmen of individual airlines). For more than two decades Mica Endsley has been researching and writing on the subject, and it is this relationship to expert performance where we find her work particularly valuable in supporting our own advocacy for the discipline of Automation Airmanship. There are by now dozens of useful models of SA, but the simplest and most practical we have found[10] (in the context of expert performance) was published in 2006; see Fig. 9-1.

We could spend several more chapters exploring the many factors that lead crews to a high level of SMA, but rather we will focus on the knowledge of SMA that experts draw on to give them *an accurate, useful mental model of relevant aircraft automated tasks, including configuration, flight and powerplant states, flight guidance, flight control and sensor modes, and their dynamic relationship to the present and future*

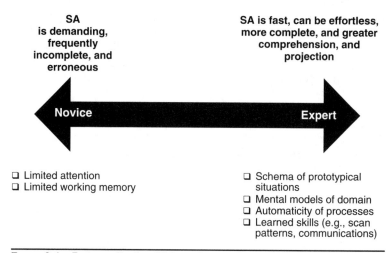

FIGURE 9-1 Factors affecting SA in novices and experts in a domain.

flight path of the aircraft. Looking at the right side of the model in Fig. 9-1, we can see a couple of valuable tools used by experts as supported by the research: comprehension, projection, and mental models. All three are constantly being updated by the very best crews, who know what a mode is, and what each mode they have selected, preselected, or armed is doing, to control the current and future flight path of the aircraft. Again, Mica Endsley comments that "Mental models and schema provide cognitive mechanisms for interpreting and projecting events in complex domains. These long-term memory structures can be used to significantly circumvent the limitations of working memory."[11]

Among the tools most commonly employed by the highest performers among us is the disciplined use of *mental models.* Because of the nature of aircraft autoflight, flight guidance, and sensor system modes, the flight deck of any highly automated aircraft is perhaps the most suitable environment in high-reliability/high-risk domains for the active use of mental models. The pathway to accurate mental models is almost always provided to every pilot in the form of the description and operational guidance provided in flight manuals and SOPs: what individual modes are, how and when they are used, what indications are associated with them, and the results they produce. Expert performers—those individuals who know the importance of accurate mental models—begin with the written guidance (what each mode is and what each mode does), then refine their understanding in training, and constantly refresh their knowledge through deliberate practice over the course of their careers.

The term *mental model* is central to our definition of SMA, just as it figures prominently in many models of SA, decision making, and other critical human factors skills. Likewise, the research community for over two decades has maintained that mental models are required to guarantee high levels of SA; industry groups such as CAST have been advocating for several years that both pilots on the flight deck be aware of ". . . the status of the guidance modes being armed or engaged and of any mode changes . . . ," and thus SMA requires a sound understanding of how mental models work in order for SMA to be effective.

A useful definition of *mental models* that has held up for several decades was provided by William Rouse and Nancy Morris[12] in 1985: "[M]ental models are the mechanisms whereby humans are able to generate descriptions of system purpose and form, explanations of system functioning and observed system states, and predictions of future system states."

Because of its rigor, we find the most useful description of how mental models factor in expert performance to be that developed by Gary Klein and his research colleagues. Our own efforts to isolate the factors that the very best crewmembers employ was in part guided by Klein's research because it so closely describes the processes used by pilots and crewmembers, working in dynamic, time-critical conditions, to achieve optimum decisions in high-stakes situations. The masthead for this chapter concisely illustrates how mental models can factor into SMA:

> Since the experts have a mental model of the task, they know how the subtasks fit together and can adapt the way they perform individual subtasks to blend in with the others. This makes their performance so smooth. They do not even feel that they are performing subtasks because the integration is so strong.[13]

In short, this describes how some pilots can appear to be so at ease with the technology on the flight deck, making their work flow effortlessly in the face of conditions that can strain average performers. "Experts," Klein says, "see inside events and objects. They have mental models of how tasks are supposed to be performed, teams are supposed to coordinate, equipment is supposed to function." Among other things that experts know are "tricks of the trade," along with the conditions for using them, and experts know enough about how their equipment works to interpret what the system is telling them.[14]

Assembling SMA on the Flight Deck

The highest-performing experts, as the researchers continue to show, conduct flight operations differently, especially with respect to the elements that comprise SMA. In addition to knowing each flight guidance and sensor mode, they know how to manage their mental model of SMA and all its dynamic components through the way that they select, preselect, and choose various modes on their cockpit displays. Using the flight guidance control panel (FGCP) or equivalent (see Table 3-1), they are constantly evaluating the configuration of the aircraft flight guidance and autoflight. In knowing this, they know what the aircraft *is doing now* and are comparing this to the aircraft clearance (or desired flight path if operating uncontrolled), and expected changes to the aircraft flight path to ensure that they have an accurate assessment of what the aircraft is going to do next. Again, Gary Klein:

> Experts distinguish themselves by their ability to anticipate what might happen next. Even while doing their work, they are positioning themselves for the next task. Their transitions from one task to the next are smooth instead of abrupt. By forming sharp expectancies, experts can notice surprises more readily. They notice novel events and the absence of expected events.[15]

Experienced, top-performing crews carefully maintain an accurate picture of both the present and future flight path states through specific, coordinated use of flight deck tools such as the FGCP, DSP, and FMS (see Table 3-1). They form "sharp expectancies" by keeping their primary flight display (PFD) and/or heads-up display (HUD) configured with only the information that contributes to SMA and supports the mental model of the how the aircraft should be acting and how it can be expected to. Knowing exactly how to configure these displays to maximize SMA takes some experience and deliberate practice, but the Checklist for Success at the end of this chapter details what experienced top performers apply routinely as they use the very powerful and reliable tools available in the advanced cockpit to support SMA.

By applying these foundational skills routinely practiced by experts, pilots can begin to approach the level of SMA that the designers, researchers, engineers, and manufacturers envisioned when they combined forces to build the most efficient and elegant family of aircraft ever operated. The safety margins that so many combined technologies are designed to increase can—if all crewmembers practice and perform as experts—be realized across all corners of aviation.

Because high-performing pilots and crewmembers have worked hard to understand the modes in which the aircraft automation operates, have practiced flying the aircraft while applying this knowledge, and have sought to gain increasingly detailed knowledge through self-critique and discussions with others (along with other techniques of self-improvement), they are able to assemble high-level automation tasks along with the subtasks. In turn, they combine this ability (MA) with the mental model that they have developed of the dynamic environment in which they are operating (SA). The process is not additive; each multiplies the other until they are, effectively, one. To blend this high-level principle with planning (the first principle), briefing and debriefing (the second principle), data entry (the third principle), communicating (the fourth principle), and monitoring (the fifth principle) is a powerful, front-line skill. Subsequent chapters that take on equally high-level principles of Automation Airmanship as SMA discuss accidents and near-accidents in which SMA (and the other five principles introduced to this point) factors heavily in the outcome, whether because of the lack of SMA, or the presence of it in the minds of the flight deck crew.

Checklist for Success: SMA

☑ Data displays reflect crew intentions during all phases of flight; the mental model of the pilot flying is compatible with the flight guidance modes, both selected (or "captured") and preselected (or "armed"). This same mental model is shared with and checked by the pilot monitoring.

☑ Cockpit displays (PFD and/or HUD) are up to date, uncluttered, and relevant to the current phase of flight. Displays are set to the appropriate level of brightness to maximize visual acuity and to support day/night vision requirements.

☑ Lateral navigation displays (ND) are kept up to date and at the appropriate range setting (at a scale compatible with the phase of flight such as cruise, arrival, approach, etc.) and with the appropriate data overlays (e.g., weather radar, terrain display).

☑ Individual spotlights are balanced over the relevant displays, so crews do not become confused by or fixated on single components of the automation.

☑ Changes in modes, when made by the pilot flying or commanded by the FMS, are confirmed by standard voice acknowledgment (annunciation of mode changes, for example) or hand signals (such as pointing to set values such as altitudes) on the appropriate display and are annunciated between both pilots.

☑ Within the highest-performing crews demonstrating high levels of SMA, automation surprises are rare; and when they occur, they are immediately resolved and debriefed as soon as practicable.

Notes and References

1. Gary Klein, *Sources of Power*. MIT Press, Cambridge, MA, 1998, p. 152.
2. The widely accepted general definition of situational awareness, or SA, is credited to researcher Mica Endsley who in 1988 described SA as "... the perception of the elements in the environment within a volume of time and space, the comprehension of their meaning and the projection of their status in the near future."
3. Nancy Leveson, L. Denise Pinnel, Sean David Sandys, Shuichi Koga, and Jon Damon Reese, "Analysing Software Specifications for Mode Confusion Potential," *Department of Computer Science and Engineering*, University of Washington, Seattle, WA, 1997, p. 3.
4. Victor Riley, "Reducing Mode Errors Through Design," *Commercial Avionics*, March 2005.
5. Anjali Joshi, Steven P. Miller, and Mats P. E. Heimdahl, "Mode Confusion Analysis of a Flight Guidance System Using Formal Methods," *Proceedings of the 22nd Digital Avionics Systems Conference (DASC '03)*, Indianapolis, Ind., Oct. 12–16, 2003.
6. See note 2., above.

7. Mike O'Leary, "Situation Awareness and Automation," British Airways Safety Services, Heathrow Airport, United Kingdom, 2000, p. 3.

8. Commercial Aviation Safety Team, Safety Enhancement 30 Revision-5, "Mode Awareness and Energy State Management Aspects of Flight Deck Automation," Final Report, August 2008, p. 8.

9. Ibid., p. 8.

10. Mica R. Endsley, "Expertise and Situation Awareness," in *The Cambridge Handbook of Expertise and Expert Performance*, K. Anders Ericsson (ed.), Cambridge University Press, Cambridge, United Kingdom, 2006, p. 637.

11. Ibid., p. 636.

12. W. B. Rouse and N. M. Morris, "On Looking into the Black Box: Prospects and Limits in the Search for Mental Models," No. DTIC #AD-A159080, Center for Man Machine Systems Research, Georgia Institute of Technology, Atlanta, 1985, p. 7.

13. Gary Klein, *Sources of Power: How People Make Decisions*, MIT Press, Cambridge, Mass., 1998, p. 152.

14. Ibid., pp. 152–153.

15. Gary Klein, *Streetlights and Shadows: Searching for the Keys to Adaptive Decision Making*, MIT Press, Cambridge, Mass., 2009. pp. 158–159.

Part **IV**

High-Level Automation Airmanship of Top-Tier Performers

Sometimes new technology—even that encouraged by law—brings with it new risks, and we are forced to confront the unthought-of consequences of a seemingly good idea.
—Henry Petroski, 2010[1]

CHAPTER 10

The Seventh Principle: Workload Management

The more complex the system, the more important this vulnerable but exceedingly flexible component, the operator, is.
—Charles Perrow, 1984[2]

This chapter brings us back to the cockpits and mission compartments of highly automated modern aircraft, where practicing Automation Airmanship principles requires more than a working knowledge of the concepts outlined up to this point in this text. Although SMA is a concept of the high art of operating a complex system in a highly dynamic, tightly coupled world, and is an essential component of expert performance, it is not practiced by experts simply for its own sake. Of the six principles (including SMA) explained thus far, not one *by itself* can guarantee safety or make an individual pilot or crew seem as expert. As is true in many occupations that demand extensive technical knowledge and skill, compliance with a wide array of regulations, operating rules, and standards along with a sound working knowledge of the human factors of complex systems is not enough. There must be a protocol for combining *all* this knowledge in a manner that is repeatable, scalable, and observable; only through this approach can the way of the expert be made into the way of all.

What has made a holistic approach to operating these complex systems difficult to achieve over the decades is that the actual *practice* of the knowledge has been fragmented into its component pieces, and the training systems (similarly fragmented) that support the operators have struggled to integrate *nontechnical skills* (such as general human factors, CRM, threat and error management, and the like) with the extensive *technical skills* required to achieve the task. In most organizations the integration of technical skills with nontechnical ("soft") skills remains clunky, at best. Experts in the operational environment, however, have been successful at doing just this, blending extensive knowledge of both the technical and nontechnical smoothly, while enjoying the high level of safety and effectiveness enabled by their advanced aircraft, just as it was envisioned by the designers and engineers. This ability in turn leads these pilots and crews to achieve the efficiencies expected by the command leadership of organizations purchasing and operating the latest designs. In Part IV of the text, which features the last of the nine principles of Automation Airmanship, we organize the first six principles into action strategies that are *repeatable, scalable, and observable*. As with all previous parts of this text, we focus on the actions of expert crews and how they demonstrate the practice of Automation Airmanship with the unencumbered elegance that you now expect to see as the standout hallmark of expert flight deck performance.

Out of Tragedy, a Clear Way Forward

In 2007 we were part of a collaboration with then Canadian Marconi Corporation (now Esterline-CMC) and the Canadian Department of National Defence (DND) to evaluate the readiness of the Canadian Air Force for what would be one of the most comprehensive fleet replenishment and replacement programs ever undertaken by a modern air force since World War II. This year-long analysis used as one measurement an application of Automation Airmanship concepts in a data collection tool specifically configured for the Canadian Air Force, to provide an objective, high-level assessment for senior command leadership. The analysis would assist the Canadian government in undertaking a comprehensive combat, transport, and search and rescue aircraft fleet replenishment. As with many organizations, the spark that ignited a bottom-up, organization-wide analysis to help guide the integration of technology into an organization widely inexperienced with such systems was not just found in the initiative and vision of leadership elements at all levels of the organization. Instead, *unfortunately* it came as part of the aftermath of the loss of an aircraft in an accident in which automation factored heavily. The story of *Tusker 914* not only sparked a model for modern Air Forces worldwide for the adoption of advanced technology, but also ignited our own passion to probe deeper into the reaches of this perennial problem common to organizations of every size. Our goal was to bring to the front lines of operational aviation a concise, comprehensive, and holistic strategy, and to make that the basis of an operational *mind-set* for crews and organizations alike.

Touted as "the world's most advanced and capable SAR helicopter available today,[3]" the Canadian Air Force derivative of the Augusta-Westland EH-101 helicopter, dubbed *Cormorant*, is indeed among the world's most advanced rotary wing aircraft, and it is far superior to its legacy predecessors in search and rescue service to the Canadian Air

Force. The integrated advanced flight deck rivals that of any comparable aircraft, placing demands on the flight crew to understand and manage complex control, flight management, and autoflight systems unparalleled by those of the steam-gauge aircraft it replaces (e.g., the Sikorsky *Sea King* and Boeing *Labrador*). It was exactly these demands that would overcome the crew of *Tusker 914* on the evening of July 13, 2006, while it was conducting training over the fishing vessel *Four Sisters No. 1* in Chedabucto Bay, Nova Scotia. The aircraft, with seven crewmembers on board, would fly into the waters of the North Atlantic at 69 knots, in a nose-low attitude, following an "autopilot-flown transition down" to a radio altitude-hold datum of 100 ft above the water (in some organizations, this maneuver is referred to as a "CATCH," or coupled approach to couple hover, although this terminology can vary across manufacturers, depending on the installed automation and the crew procedures). The survivors included two pilots seated in the cockpit, a third pilot seated in the cockpit jump seat (the aircraft captain) and a search and rescue technician, or SAR Tech. Three crewmembers in the aft cabin of the helicopter would lose their lives as a result of the crash, whose repercussions would be felt for years as the Canadian Air Force would go on to adopt automation protocols and other flight standards policies which would place them in the forefront of aviation organizations worldwide.

There are many reasons why regulators, manufacturers, operators, and the representatives of survivors and victims of accidents search exhaustively for meaning in the sullen aftermath of an accident. Among the most important reasons is to reach a conclusive understanding of how such a heavily defended and carefully designed system can succumb to internal and external dynamics that can—seemingly unknown to the humans operating it—tear the fabric of such systems apart, either suddenly with violence or gradually, fiber by fiber. Not surprisingly, investigators normally reveal many factors at work as they go about their careful business; sometimes they can find, in that last moment beyond which an accident could be avoided, a pathway for escaping imminent disaster. Once that pathway is identified, the challenge becomes one of transforming the cold evidence and hard facts into knowledge and actions that can guide other crews in other, similar situations along the path to mission completion, instead of the path to disaster. The legacy of *Tusker 914*, with its many complicated dynamics, is one that promotes the understanding of the interaction between the comparatively flexible human component and the complex yet predictably dutiful technology. But before we can understand the lessons of this fatal mission, we must return to the qualities that distinguish the components of the human-machine partnership. It is this understanding, in turn, that helps us to build a reliable model that can guide operators toward optimum performance in the high-stakes environment of modern aviation.

A Natural Division of Labor between the Wetware and the Machine

For several decades, experts in human factors and aviation psychology have offered a few useful models intended to help pilots and crews organize their interaction with complex automation. Several of these models have been used successfully in organizations and have even become institutionalized in regulations and SOPs. After decades of working with these models, we fervently believe that the goal of such assistants should be to create a balance between flight deck workload and SMA that is optimized for a specific phase of flight and is supported by the appropriate "level" of automatic control over the aircraft flight path. We can get to this high level of understanding of today's complex systems by examining the basic principles that cause engineers and designers to allocate some tasks to the human operators and others to automation.

In the early 1950s the research of P. M. Fitts (one of the pioneers of human factors in aviation) led to what has since been commonly referred to as the *Fitts list*.[4] For over 60 years it has endured, and it remains remarkably resilient even in the face of ever-increasing interest by other human factors researchers around the world. If you have ever driven an automobile, used a smart-phone, adapted a personal computer or other electronic device to your lifestyle, much less flown any advanced aircraft, you will immediately recognize the usefulness of the Fitts list (Table 10-1).

Since the Fitts list was introduced in the early 1950s, there have been significant advances by "machines" onto ground that was once held exclusively by humans, most notably long-term memory (principally storage and retrieval of information) and sensory functions (just think of the advances in radar technology alone!). Some 60 years later, we are more ready to hand over functions to automation that would have been more jealously guarded just a few decades ago (think of how the arrival of TCAS in the early 1990s changed how pilots handled traffic conflicts!). Every time a pilot

Humans Are Good at:	Machines Are Good at:
1. Sensory functions, detection	1. Speed and power
2. Perceptual ability	2. Routine, repetitive work
3. Flexibility, improvisation	3. Computation
4. Judgment and selective recall	4. Short-term storage
5. Reasoning	5. Simultaneous operations
6. Long-term memory	6. Short-term memory

TABLE 10-1   The Fitts List

engages an autopilot or autothrottle, activates a new flight plan or waypoint string in the FMS, queues a data-link message for downlink or any one of countless related requests of the machine, he or she is relying on the inherent ability of the automation for those tasks to which automation is best suited. But doing so requires more than casually conceding routine and repetitive tasks to the tireless reliability of the machine; it requires *actively exercising those specific duties for which a reliable substitute has not yet been devised: the operator.* The duties that cannot be delegated to the automation include flexibility, improvisation, judgment, reasoning, and short- and long-term memory of patterns and situations that help the human operator maintain safe control of the flight path vector. Former NASA research engineer Prof. Steven J. Landry of Purdue University writes of the contemporary interpretation of the Fitts list:

> In general, it appears humans are 'wired' for rapid operation in a highly uncertain and diverse environment, whereas machines are still generally relegated to rather specific operations in a well-regulated and defined environment. Therefore, while machines may be better at routine and repetitive work, they are generally unable to deal with events or circumstances outside of the expected operating regime. Humans, however, are able to operate in the face of such events, leading to the notion that humans are suited for "supervising" automation.[5]

The Human Operator as Ultimate Authority

We would add to Landry's assessment an additional notion: that humans are best suited—through an understanding of Automation Airmanship—for *taking over control* for automation whenever their judgment and experience indicate that the automation is not appropriate for the moment, or is not behaving as planned.

Exactly when and how to "take over" for the automation continues to vex many operators in modern aviation, a problem that played a prominent role for the crew of *Tusker 914.* As the helicopter approached its target altitude of 100 ft above Chedabucto Bay for its rendezvous with the *Four Sisters,* the first officer (FO) flying the aircraft was supervising the autoflight system in a normal 7° nose-up attitude. According to SOP, the FO transferred control of the aircraft to the other pilot (under supervision of the aircraft captain in the jump seat, the pilot flying was designated the acting aircraft captain (AAC). Shortly after the transfer of aircraft control, the AAC began to apply control forces to the aircraft controls (primarily the cyclic, or "stick," with the autopilot still engaged), resulting in a climb to 170 ft radio altitude and an acceleration to 30 knots. At this point, as many autoflight systems are designed to do in similar situations, the AAC's control inputs ceased, and the autopilot resumed control toward its active mode, a slow-speed forward hover, but at the new datum of 170 ft above the water, as a result of the overriding inputs of the AAC. With the AAC recognizing the inadvertent climb, what followed was a series of pitch and roll inputs by the AAC, again overriding (but not disconnecting) the autoflight, each input, however slight, successively demanding more counteracting force by the autoflight system to maintain or regain shifting targets associated with its current flight mode: *slow-speed, forward hover.*

With the situation deteriorating and the aircraft descending below 85 ft above the water (and the "Check height!" aural warning sounding), the aircraft captain in the jump seat called out, "Go around, go around, go around!" Still overriding (but not disconnecting) the autopilot with manual control inputs, the AAC announced that he was going around, and a new autoflight mode was selected by the crew, *Transition Up.* With the *autopilot still engaged,* a new mode selected by the crew, and continued manual inputs to the flight controls coming from the AAC, the system (specifically, the pitch series actuator) became so saturated that when the controls were momentarily released by the pilot in a 20° nose-down attitude, the system could not fully recover to accomplish the commanded transition-up maneuver. Seconds later, the aircraft impacted the water 18° nose-down, and with the autoflight system still attempting to recover to its commanded flight path. The official accident report would subsequently conclude that ". . . the helicopter responded normally to manual and AFCS [automatic flight control system] inputs and there was no evidence that a system malfunction contributed to the accident."[6]

Balancing Responsibilities between the Wetware and the Machine

Few things in aviation are as complex as the flight control and autoflight systems of an advanced helicopter. It is widely accepted that these systems require a deeper knowledge by their crews than comparable but less dynamic systems found in most fixed-wing aircraft. Even so, the allocation of functions between the pilots and the automated systems is fundamentally the same: the crew delegates certain functions to the aircraft to balance workload, optimize SMA, and maintain proficiency. Modern cockpits are designed to offer a multitude of choices for how much automation can be used for any phase of flight—in some cases, it is not only obvious, but also recommended by policy or regulation. A Category III autoland, for instance, generally requires that the autoflight be fully coupled to the autopilot and autothrottles; operations in RVSM airspace requires autopilots. There are many more situations in which it is impractical to mandate what tasks are accomplished by the autoflight, leaving pilots the flexibility to decide for themselves, again balancing workload, optimizing SMA, and maintaining proficiency. The model we advocate is based on these very simple concepts, yet we believe it is durable enough to be applied to any cockpit and design or SOP (Fig. 10-1).

There are many ways to represent the various levels of automation that are available to the flight crew for every design of advanced cockpit, and every manufacturer will have its unique requirements that can impact the model we advocate below. We call it the Balanced Model® for levels of automation because it requires that crews *work together* to balance *both SMA and workload* at every level, keeping the level of autoflight appropriate to safely control the aircraft flight path

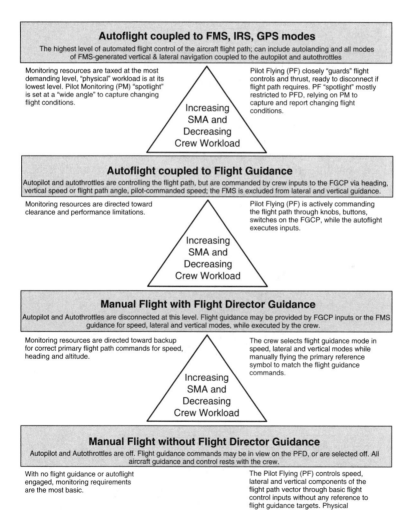

Autoflight coupled to FMS, IRS, GPS modes

The highest level of automated flight control of the aircraft flight path; can include autolanding and all modes of FMS-generated vertical & lateral navigation coupled to the autopilot and autothrottles

Monitoring resources are taxed at the most demanding level, "physical" workload is at its lowest level. Pilot Monitoring (PM) "spotlight" is set at a "wide angle" to capture changing flight conditions.

Pilot Flying (PF) closely "guards" flight controls and thrust, ready to disconnect if flight path requires. PF "spotlight" mostly restricted to PFD, relying on PM to capture and report changing flight conditions.

Increasing SMA and Decreasing Crew Workload

Autoflight coupled to Flight Guidance

Autopilot and autothrottles are controlling the flight path, but are commanded by crew inputs to the FGCP via heading, vertical speed or flight path angle, pilot-commanded speed; the FMS is excluded from lateral and vertical guidance.

Monitoring resources are directed toward clearance and performance limitations.

Pilot Flying (PF) is actively commanding the flight path through knobs, buttons, switches on the FGCP, while the autoflight executes inputs.

Increasing SMA and Decreasing Crew Workload

Manual Flight with Flight Director Guidance

Autopilot and Autothrottles are disconnected at this level. Flight guidance may be provided by FGCP inputs or the FMS guidance for speed, lateral and vertical modes, while executed by the crew.

Monitoring resources are directed toward backup for correct primary flight path commands for speed, heading and altitude.

The crew selects flight guidance mode in speed, lateral and vertical modes while manually flying the primary reference symbol to match the flight guidance commands.

Increasing SMA and Decreasing Crew Workload

Manual Flight without Flight Director Guidance

Autopilot and Autothrottles are off. Flight guidance commands may be in view on the PFD, or are selected off. All aircraft guidance and control rests with the crew.

With no flight guidance or autoflight engaged, monitoring requirements are the most basic.

The Pilot Flying (PF) controls speed, lateral and vertical components of the flight path vector through basic flight control inputs without any reference to flight guidance targets. Physical workload is at its highest.

FIGURE 10-1 The Balanced Model for levels of automation.

desired by the crew. It is simple: when crew-desired flight path differs from what is being commanded or controlled by autoflight or flight guidance, reverting to a lower level makes the most sense, from the standpoint of both SMA and workload. *When this relationship is out of balance, select autoflight systems OFF, reconfigure or disregard autoflight, and adjust the monitoring spotlight of both crewmembers for the new level.* As balance is achieved at each level, a higher level can be engaged, if desired by the crew. As a rule, SMA should increase with each level, and physical workload should go down. An out-of-balance condition might look like Fig. 10-2.

Experienced top performers do not fly all the time in perfect balance —flight conditions can change rapidly for countless reasons— but they do demonstrate the ability to move up and down, from level to level (even skipping a level, or two!) with ease and smoothness. Most of all, they know why they are at each level and how this affects their SMA and workload. What is more, experts know exactly how to move between levels with minimum effort, and without aural and visual warnings creating unnecessary distractions for the entire crew:

> An expert anticipates that after seeing the approach lights and runway environment at the completion of a nonprecision approach, he will smoothly disconnect the autopilot (and smoothly cancel and silence any warnings), and command the PM to set the desired

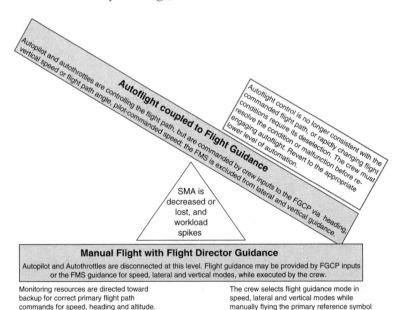

FIGURE 10-2 Automation level out of balance with SMA and workload.

vertical flight guidance, visually maneuvering the aircraft with ease (manually) to the extended runway centerline, while listening carefully to the PM providing sink rates, height above the runway and airspeed.

In a similar fashion,

> An expert experiences an in-flight upset, encountering severe turbulence while at high altitude, and disconnects the autopilot and autothrottles, sets constant thrust and pitch, smoothly cancels and silences disconnect warnings, and directs her or his spotlight to the primary flight parameters displayed on the PFD. With a balanced level of automation restored, the PF calls for the appropriate emergency or abnormal procedures.

These examples demonstrate expert crews in situations that were in one instance anticipated and in the other instance a complete surprise. Similar instances occur across the globe countless times each day, with crews seeking out the appropriate balance between SMA and workload, consistent with monitoring resources, to protect the integrity of the aircraft's flight path. These crews are, whether conscious of the fact or not, actively practicing the Automation Airmanship seventh principle.

Looking back on the sequence of events leading up to the crash of *Tusker 914*, it is easy (as it always seems to be, after the fact) to see where this crew could have applied the principle of workload management to change the course of events. Operating at the highest level (autopilot and autothrottles coupled to flight guidance) as they approached their target altitude and position over the *Four Sisters No. 1*, the crew's first chance came when the aircraft began to climb and accelerate. From this point until just a few seconds before impact, there were other opportunities, but even after the "Go around, go around, go around!" command given by the aircraft captain, the autoflight remained engaged; not only were the automated systems of the *Cormorant* saturated, so too was the crew's workload level.

Proficiency at Every Level

Inevitably we come to a common topic in active debate throughout the industry, brought to focus by recent, high-profile accidents, of the appropriate level of proficiency that pilots should demonstrate at the lowest level of automation: flying without any automation at all. We agree with many organizations that have adopted the credo "pilots will be proficient at all levels of automation" and advocate that approach, which seems clear enough. Executing this guidance can be somewhat less clear, however, as many pilots complain that their colleagues are increasingly less proficient at *manual flying*. Appropriate practice is called for, at every level, and organizations must confront this requirement resolutely so that flight crews can smoothly move between levels of automation, whether anticipated or not. Any policy adopted, we feel, must be infused with the knowledge of how automated systems are designed to handle these transitions, enabling smooth control by the crew at all levels of automation. We would like to offer the approach that we have seen many experienced top performers demonstrate successfully, and provide insight to individual pilots on how they can achieve the same level of performance across the seventh principle as so many experts.

A common mistake when flying automated aircraft is that when the automation is intentionally disengaged, such as when preparing to land, the pilot flying immediately begins to make flight control and throttle inputs. When the automation is disengaged in this situation—*in most normal instances*—the aircraft has been properly trimmed *by the autoflight system,* and the powerplant control has likely left the aircraft *on speed.* Thus any inputs from the pilot are likely *unnecessary* and will most probably lead to the aircraft deviating from the desired flight path and airspeed. Here is where effective monitoring is blended smoothly with all other flying skills: expert pilots are never in a rush to make inputs and/or corrections. Inputs and corrections should be made only if the aircraft airspeed, attitude, heading, and altitude are deviating from desired. These experts are known among their peers as "smooth" pilots or "low-gain" pilots because they never seem to be in a rush to make corrections, yet they always seem to be on airspeed and altitude. They are like a damped system whose inputs become smaller and smaller as the system approaches the desired state, and they are able to maintain that state with minimal inputs. Often the *only* indication to fellow crewmembers that a low-gain pilot has disconnected the autoflight is the actual autopilot/autothrottle disconnect aural and visual warnings (which are also silenced with ease and fluid action): *All the time in the world.* Indeed.

Pilots that immediately begin to make flight control and airspeed inputs after automation is intentionally disengaged are described as "rough" pilots by their peers, or "high-gain" pilots. They seem to always be making corrections, even when corrections are not called for. They usually are able to maintain airspeed and altitude; however, any little distraction or disturbance can cause them to begin to deviate further and further from target flight path parameters until dramatic intervention or disaster occurs. They are like an undamped system where inputs become larger and larger and the system never approaches the desired state, or if it does, it quickly deviates away again. Even when maintaining the desired state, these pilots use large inputs that are unnecessary; for these pilots, there *never seems to be enough time* for making corrective inputs.

We understand that we have described the situation where a crew anticipates the disengagement of the autopilot, autothrottles, or both.

But what course of action can we reasonably expect when unanticipated conditions result in a sudden (and often dramatic) disconnect of autoflight? Obviously if the automation disengages due to a malfunction or failure, immediate pilot input is most likely required (aircraft out of trim, powerplant failures or malfunctions, etc.). Again, practice in this area should be frequent and deliberately evaluated (optimally, in full-motion simulators). This kind of practice will likely create pilots who maintain their normal basic airmanship scan during autoflight engagement or disengagement, until a smooth handoff has occurred. The results of this kind of training are pilots who can quickly ascertain what inputs are required to return the aircraft to the proper airspeed, attitude, heading, and altitude. It sounds simple enough; but our industry continues to accumulate accidents and incidents related to the improper handling of the aircraft, at low altitudes, middle altitudes, and high altitudes, in spite of the technology installed on the flight deck to prevent such flight path excursions. Defending against these occurrences has been the topic of this chapter. Chapter 11 presents the eighth principle, *positive flight path control*, as a way to recover from unwanted flight path excursions when they occur.

Checklist for Success: Workload Management

☑ Use automatic control primarily to balance workload with SMA, keeping the level of automation appropriate to the phase of flight and desired workload for each crewmember.

☑ Automatic and/or crew-assisted transitions between flight modes do not occur without the consent and knowledge of all crewmembers.

☑ Intervene immediately when automation fails to control the aircraft along the intended flight path; reduce the level of automated control or disconnect automatic control completely.

☑ Without compromising positive control, make transitions between levels of automation smoothly, including the cancelling or silencing of backup warnings and alerts associated with autoflight disconnects.

☑ After moving from one level of automation to the next—— adjust monitoring of the aircraft flight path consistent with required crew workload.

Notes and References

1. Henry Petroski, *The Essential Engineer: Why Science Alone Will Not Solve Our Global Problems*. Alfred A. Knopf. New York, NY. 2010. p. 6.
2. Charles Perrow, *Normal Accidents: Living with High Risk Technologies*, Princeton University Press, Princeton, NJ. 1999. p. 271.

3. www.agustawestland.com
4. Steven J. Landry, Human-Computer Interaction in Aerospace. In The Human-Computer Interaction Handbook, 2d Ed. Andrew Sears, Julie A. Jacko, eds. Taylor and Francis Croup, New York, NY, 2008, p. 727.
5. Ibid.
6. Canadian Forces Flight Safety Investigation Report (FSIR), No. 1010-149914 (DFS 2-3), January 22, 2008, pp. 34–37.

CHAPTER 11

The Eighth Principle: Positive Flight Path Control

Fly first.
—David Owens and Michael Varney, 2008[1]

There is a frontier beyond excellent planning and expert briefing, errorfree data entry and effective communications, knowledge-based monitoring, accurate situational and mode awareness (SMA), and a balanced use of automation that must be crossed into eventually by every glass cockpit pilot, whether in actual operational experience or in regular, highly realistic flight training. The frontier we are referring to is increasingly gaining notoriety in our industry and remains shrouded in ambiguity, indecision, and competing opinions; it is known simply as *failure*. Failure in this context is inclusive of individual aircraft components (including hardware and software); an individual organization's processes, procedures, and practices; the increasingly complex global air traffic control and regulatory system; the dynamics of high-performance teams; and of course, the complexities of the individual. This chapter targets the role of automation and the crew *in the realm of failure* and how experts form durable strategies not only to avoid failure, but also to confront it when it occurs.

It is impossible to predict where and when the next disaster (or near-accident) will happen, but it seems inevitable that the failure of automation to behave *as expected* or *as designed* will continue to provide both drama for the media outlets to report on and events for the accident investigators to analyze for many more decades. We feel strongly that those of us who work in the field every day can take immediate and fundamental action to significantly reduce these occurrences. Up to this point, we have omitted any discussion of the two events that have dominated aviation safety debates during most of the writing of this book. The resulting revelations that have come from the official investigative organizations in both accidents have reverberated through industry trade groups; safety and regulatory organizations; global, regional, and local flight departments and airlines; and even major manufacturers and aircraft flight deck designers. The drama of both Air France flight 447 and Colgan Air 3407 continues to build at the time of this writing, as do the many opinions of what can be done to avert such avoidable accidents in the future, seemingly setting the aviation safety agenda for the next decade. The first, an Airbus A330, crashed into the Atlantic in June 2009 while en route from Rio de Janeiro to Paris, killing all 228 persons on board; the second, a DHC-8-400 that crashed near Buffalo, New York, in February 2009 killed all 49 on board and 1 person on the ground.

We have avoided discussion of both accidents up to this point so that we could evaluate them in the context of the information presented in the previous 10 chapters and thus minimize our contribution to the cacophony of critics which builds almost daily around these two accident flights. Both accidents have raised the stakes for the industry in its effort to prevent or recover from unexpected in-flight events that appear to be easily overcome if only the crew had executed a few basic steps of fundamental airmanship. In our analysis, both events dramatically demonstrate the distance that separates expert performance from inappropriate flight deck behavior. Because both accidents involve complex, glass cockpit aircraft, we must include them in our discussions. But in our effort to balance the examination of the incidents across our industry that point to a lack of skill, we will also report in this chapter on incidents with successful outcomes that are widely underreported compared to Air France 447 and Colgan Air 3407, but that actually hold more clues to repeatable, reliable, and safer outcomes than those of comparatively more dramatic disasters.

From the opening page of this text we have examined crews of modern aircraft as they respond in the face of uncertain information, or failure of automation to behave as expected, and we have witnessed the dramatic results that can occur when the failure sequence overwhelms the ability of the crew to recover. For our part, we have seen countless crews demonstrate exceptional Automation Airmanship and have learned much from this valuable experience. These expert crewmembers have honed a sharp performance edge out of the knowledge of their own performance limitations, their aircraft's automated alerting and warning systems, familiarity with their supporting flight documents (such as checklists and SOPs), and the fundamental approach to abnormal events that begins with controlling and stabilizing the flight path of the airplane *before* calling for the checklist or troubleshooting "root causes." As a result, their performance, repeatable over a wide variety of abnormal events or emergencies, flows seemingly effortlessly with every challenge they face. It is time to make what they do and how they do it known to everyone; it is *past* time to demonstrate the mastery potential that every pilot and crewmember should be able to demonstrate on short notice, time after time, across generations and in situations as different as the long-range cruise leg of a transoceanic flight and a night-time approach in icing conditions.

How the Wetware Responds to the Unexpected

Once more, we think that a sound discipline based upon knowledge, refined in practice, and supported operationally is the best way for an organization to inoculate itself *not against the eventuality of failure*, but from the disaster that can result. We begin this discussion, as we have several others, with a concise summary of the human component, or the wetware, and how experienced top performers understand and act on the knowledge of this critically important flight deck component. As in previous chapters, we aim to tighten the integration of the human operator with the advanced technology flight deck by providing the foundational knowledge that experts have acquired over their careers and that others can apply just as the experts do.

One of the most persistent models for human performance under stress actually dates to 1908, but continues to be taught in a variety of disciplines related to human performance as it continues to be validated by the research community in a variety of ways. The *Yerkes-Dodson law*,[2] also known as the *inverted U* or the *stress-performance curve*, provides a working model that is universally understood as the simplest way to describe human performance when arousal or stress[3] is increased (Fig. 11-1). At the low end of the curve, where the challenge of a situation is minimal, so is performance; but as stress is increased to an "optimum" level, performance is at its peak. Beyond this point, for both simple tasks and complex tasks alike, performance degrades. Automation researchers Chris Wickens and Justin Hollands further point out that ". . . the 'knee' of the curve, or the optimum level of arousal, is at a lower level for the more complex task (or [for] the less skilled operator) than for the simpler task (or [for the] expert operator)." In other words, confirming what we all sense when in the presence of expert performance, that which seems difficult to less experienced crewmembers is made to look easy by experts, *even under stress*. We know that this graphic illustration is simple and that it does not represent actual data gathered in advanced cockpits by experienced and inexperienced crews in a controlled study of any kind. We are not presenting ground-breaking research with this model, merely a generalized visual model that we think works for practical pilots on the working advanced flight deck where the model explains in the simplest terms the relationship between stress and performance (see Fig. 11-1).

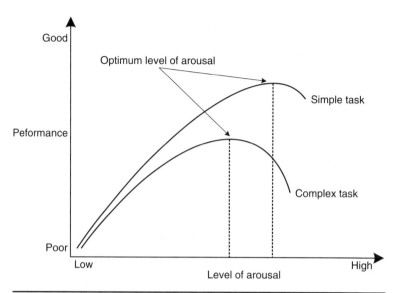

Figure 11-1 The Yerkes-Dodson law.

Admittedly, most readers are probably familiar with this popular model and are also likely to draw parallels between this model and the "spotlight" and "conveyor belt" concepts discussed in Chap. 8, The Fifth Principle: Monitoring. While we think that the research community could provide much insight into flight crew performance at the peak—and beyond the peak—of the stress-performance curve, experienced top performers seem to go about their cockpit activities in a way that does not result in a marked decrease in performance during difficult situations. Whether it results from an understanding of this specific model of stress and performance or some other reliable strategy for limiting the effects of stress, experienced top performers are able to bring this knowledge into practical use across the spectrum of failures, emergencies, and unanticipated events such as in-flight upsets. These experts know that they are limited, physically and cognitively, in their ability to execute tasks with ease, insight, and creativity, when they find themselves on the descending portion of the inverted U. It normally looks like this: in the face of a demanding, time-critical situation in flight, experts are able to "apply the brakes" and limit the multitude of options available to them to the few actions that will stabilize the developing threat and prevent it from overwhelming the crew and the available resources. Often this is accompanied by reducing the amount of automation in use, reverting to manual control of the flight path, and managing the shared workload among crewmembers as evenly as possible. Other strategies, as simple as they sound, are just as reliable: advising air traffic control to "stand by"; requesting a vector, holding pattern, or 360° turn; and, as obvious as it sounds, turning on the cockpit lights during the execution of an emergency checklist procedure at night or in conditions of low light.

Experts know how they think in demanding situations on the flight deck, and they use this powerful knowledge as a performance advantage when unexpected events occur. Increasingly we find experts who are in tune with the advances made by researchers in the field of cognitive neuroscience, and we are encouraged that more and more training programs are providing pilots and crews with knowledge of this important multiplier of personal performance. In Chap. 8 we noted that the very best performers across all high-risk/high-reliability domains are "thinking about their own thinking" and *that very strategy* can have a dramatic impact on performance when crews crest the top of the curve and find themselves teetering toward the steep downslope beyond the summit. It is no surprise, then, that as we look across the research of the past decade, we find evidence that adds to the knowledge of the unique abilities of experts to handle difficult situations. Drawing from their research into high-performance teams (conducted with an emphasis on how communications can play a role in managing stress), Eduardo Salas and his colleagues conclude that

> A challenge in decision making within team environments occurs when team members begin to experience stress and can't easily diagnose the situation because of the performance-degrading effects of stress. Normally higher-status members are less likely to take the advice of lower-status members. However, under stress three things happen. First, higher-status members are more willing to accept input from those with less expertise, but under these conditions low-status members generally aren't vocal in their viewpoints. Second, attention tends to narrow, producing a form of tunnel vision. Third, explicit communication decreases as members become more focused on their own respective roles. Despite these challenges expert teams are adaptive and able to maintain coordination levels and corresponding effective decisions despite these conditions; *they have behavioral and cognitive mechanisms in place that allow them to maintain high levels of performance* [emphasis added].[4]

Some of these peak performers have pursued their knowledge of the wetware and are aware of the physiology of stress which, when added to their experience and knowledge, form a powerful combination in the face of stressful, in-flight emergencies. We know of several training programs that have been advocating for over a decade that flight crews learn and apply the physiological fundamentals of stress. On the increasingly complex flight deck often operated by the minimum required crew, we think this is another component of "essential systems knowledge" for contemporary flight crews. This lesson might include the fact that it is the brain's amygdala that detects danger and sends messages to the hypothalamus, which sends a message to the pituitary gland that in turn releases adrenocorticotropic hormone, or ACTH. ACTH acts on the adrenal cortex, and steroid hormones (also known as corticosterone or CORT) are released into the bloodstream, a process that repeats until the event stimulating the sequence is past.[5] There are dozens of physiological outcomes from this process, from the intensifying of memory during such events to the physical breakdowns that result from chronic exposure to stress. Expert glass cockpit crewmembers may not be able to recite the "stress pathway" from memory while under stress, but the one thing that top performers (who know the basics of stress) can count on is that their ability to think creatively, to carefully consider options, to reason through a complex solution and then to devise a careful plan of execution is reduced dramatically under stress. And in this case the very best do those few things *that they know from experience* will stabilize the situation and return it to a state where they can apply their higher-order mental processes to the problem.

There are a variety of terms associated with acute, situational stress not unlike the 4 minutes 23 seconds of harrowing descent that the crew of AF 447 endured in the darkness over the Atlantic Ocean in 2009,[6] or the final 27 seconds of Colgan Air 3407 near Buffalo, New York, earlier that same year.[7] *Channelization, tunnel vision,* and *downshifting* are just several terms to describe what happens to the brain during emergency situations that are sudden, unfamiliar, and unexpected. A number of studies have shown this very effect with convincing results: both novel and creative behaviors are disrupted, while individuals are likely to continue with a given action or plan of action that they have used in the past, even if it is ineffective (a concept known as *perseveration*).[8] Again, Wickens and Hollands comment that

> Stress will initially narrow the set of cues processed to those that are perceived to be most important; as these cues are viewed to support one hypothesis, the decision maker will perseverate to consider only that hypothesis and will process the (restricted) range of cues consistent with that set. That is, stress will enhance . . . confirmation bias causing the decision maker to be even less likely to consider the information (process cues) that might support an alternative hypothesis.[9]

The Pathway to Recovery from Failure and Surprise (the Wetware)

This describes the behavior of many mishap crews, as evidenced by cockpit voice recorder (CVR) transcripts and digital flight data recorder (DFDR) data gathered from accident flights for the past several decades; but it does not explain how some crews are able to persist in the face of physiological limits and emerge from similar situations safely and unharmed. Experts who experience similar events but with different outcomes know their personal limits; but more importantly they know what to do when their limits are exceeded due to circumstances or time lines that are beyond their control. The researchers have provided ample evidence in favor of at least four adaptive strategies that "skilled operators" demonstrate.[10] One strategy, *recruitment of more resources* (trying harder, or getting help), is common and has risks associated with it, primarily a "shift in the speed-accuracy tradeoff" where the situation can be responded to quickly, but with errors that must be resolved after recovery. Calling another crewmember to the flight deck, for example, is a long-term strategy that is not very practical when the aircraft's flight path is out of the boundaries of normal performance. Another strategy, *removing the stressor*, is often impractical in flight, unless the time pressure is not a factor; unfortunately, most accidents contain some element of time pressure, and every flight must eventually end as fuel is exhausted. A third strategy, *do nothing*, in most emergency situations is entirely impractical—not many failures heal themselves by being left alone. But one additional strategy, *changing the goals of the task,* actually has shown significant advantage as the most viable strategy in time-compressed situations. Recent research has proved the practicality of individuals accepting simpler solutions to problems that require less working memory and have been labeled as "more stress resistant."[11] And finally, in discussions in the research community on the subject of stress and performance, research has shown that highly skilled operators are generally more "immune" or "buffered" from the negative effects of stress than their less-experienced colleagues. Three reasons are cited for this:

1. Skill leads to automaticity: knowledge-based behavior is replaced by rule- and skill-based behavior.

2. Experts have a greater repertoire of strategies available to perform a given task.

3. Higher levels of experience can lead to greater familiarity with stress and with its effects on an individual's performance.

It seems like a lot for experts to know. But somehow, they do. Talk to a group of experts, or any widely respected and skilled colleagues, and they will reveal patterns of behavior, knowledge of themselves, and *how they think* when inside the borders of stressful situations. In essence, if not in concise technical terms, experts can describe, when asked, the concepts that the research community has been proving now for many decades. Yet still there is more: the very best experts in the field, across several high-risk/high-reliability domains, have been adopting the knowledge of the human operator that the research community continues to develop every day. They are keeping up with the research into human performance just as they also keep up with their knowledge of the technical environment they work in.

Recall that, in the opening paragraph of this chapter, we said the failure we are contemplating involves *all the components* on the flight deck, from the hardware and software to the wetware operating it. A crewmember's response to a situation where he or she perceives that failure to be with the aircraft can be expected to be handled *in the same way as if the failure were in the individual and/or the crew.* In both the AF 447 and Colgan Air 3407 accidents, the engines and primary flight controls of both aircraft (according to the available evidence at the time of this writing) were determined by investigators to be functioning normally.[12] There remains substantial investigation to be accomplished (again, at the time of this writing) into the possible role of AF 447's instrumentation and sensor systems prior to and during the A330's 4-minute plummet into the sea; but in the case of Flight 3407, the investigation isolated the cause solely to the crew's actions. In summary, both aircraft were flyable, and both aircraft should have been recovered without a heroic feat of airmanship, but rather with basic, *unheroic* flying that in hindsight would have been viewed as quite routine. Naturally, by looking backward through the lens of hindsight, at "ground speed zero, 1.0*g*" and with the cool facts provided by CVR transcripts and DFDR data, it is easily concluded that both crews behaved inappropriately, while there are many details of both accidents that combined to lead both crews to act as they did. But as we have discussed, it is possible that crews might *respond* to an alert or warning as if it were a failure of the aircraft, and not attributable to pilot action. We suggest that regardless of how these crews or others find themselves suddenly on the downslope of the inverted U, *the pathway back to a state of optimum performance is the same.* The researchers have told us that it is a fool's errand to even attempt to resolve the "why" or "how" of a bad situation that is rapidly becoming worse when there are much simpler and much more important things to do. How do the experts do it—what do they know about their aircraft, in addition to what they know about themselves, that separates them from average performers? It is another area of research that might well change the way that contemporary flight crews are trained. But in our observations and our careful work with experienced instructors and evaluators across a wide variety of aircraft and missions, we suggest that all top-performing aviators learn to effectively blend the "systems knowledge" of both themselves and their aircraft in the way that all expert performers do, forging a powerful combination of knowledge that both makes them resistant to the steep slope of failure and serves as a kind of "emergency brake," slowing or even reversing the descent and pushing them back to their cognitive center, where they know they perform best.

Hardware and Software under Failure

Pilots who are either just beginning to learn a new advanced aircraft or who are continuing their learning while accumulating years of experience have little say over the makeup of the technology on the flight deck or mission compartments that they interface with as crewmembers. Very few pilots have the resources to choose their favorite combination of FMS, instrumentation, sensor systems, and cockpit layouts as well as the multitude of support equipment, information, and training that comes with any advanced aircraft. Virtually all pilots must adapt to their surroundings pretty much as they find them, relying on training and other supporting infrastructure (such as flight handbooks, checklists, and other "interfaces," for example, quick reference handbooks or electronic checklists that are integrated directly into the flight deck) to provide them with the knowledge necessary to operate with the highest level of safety and operational efficiency. The very best crewmembers know that it is impossible to prepare for every possible outcome, and so they seek to develop a discipline that is adaptable to every situation, no matter how complicated. These individuals have usually gone beyond the scope of most training programs to obtain knowledge on their own of how systems fail, whether the system is simple or complex.

In his book *Normal Accidents*, the author and researcher Charles Perrow provides foundational knowledge of how accidents can occur in both simple and complex systems, of which many combine to create the modern flight deck. In his analysis he examines interactions between systems, describing the possibilities as either *linear* or *complex*. Linear interactions are the most common and are described as visible, relatively simple to understand, and predictable. They also have minimal feedback loops and lack specialization. Complex interactions are not as common, are not usually visible, and are not immediately comprehensible. These interactions have multiple feedback loops and require specialization. This approach to understanding systems can be easily applied to unexpected events on the flight deck, from the well-practiced critical immediate-action steps of an engine failure at "Vee-one"[13] to the comparatively complicated and difficult-to-resolve abnormal/emergency known as *lost, suspect, or erratic airspeed*. The former often is characterized by unmistakable cues and requires near-immediate recognition and response, while the latter can develop over many minutes, often is confusing and ambiguous, and demands careful, deliberate analysis for safe, accurate resolution.[14]

Another dimension of interacting systems is the concept of coupling, which can be broken down further into *tightly coupled* and *loosely coupled*. Tightly coupled systems offer no "slack" or "buffer" between two items (such as steps in a checklist, or the action and reaction to a flight control input on behalf of the pilot). Tightly coupled systems are inherently unforgiving, time-dependent, and inflexible. Conversely, loosely coupled systems tend to have "ambiguous" or "flexible" performance standards (as in the relationship between how precisely an arrival and approach are programmed, planned, and briefed by the crew compared to the variety of ways in which it can be successfully flown, and the alternative combinations available for configurations, airspeeds, and descent profiles along the flight path). Loosely coupled systems are tolerant of delays, incomplete steps, interchangeable resources, substitution, and waste.[15]

In recovery from failure, *redundancies, buffers, and substitutions* must be thought of in advance and designed into the process or procedures in tightly coupled systems. Equipment, automation, procedures, and training are some ways in which tightly coupled systems can be protected from catastrophe when failure confronts the crew (much evidence supports this notion). In loosely coupled systems, delays, consideration of alternatives, and even improvisation may contribute to the successful outcome (and there is much evidence of this as well). Experienced top performers may not be students of Perrow, but their actions demonstrate a basic understanding of the nature of both systems, and they understand when it is essential to execute critical action steps in a procedure from memories created by deliberate practice. Conversely, they know when it is more prudent to proceed cautiously, compare the outcomes created by different courses of action, and incorporate other resources that more time-critical failures do not tolerate. What is more, with these top performers, they know how to prevent a linear system from becoming complex, and they do not allow loosely coupled systems to become a tightly coupled system that may demand time-critical responses to recover successfully. We think that more research *and greater practice* of these types of situations would increase the resiliency of crews in situations where they occur in operational flying.

One of the most dramatic advances in aircraft design in recent decades has been the amount of information that onboard system monitors are capable of handling, organizing, and displaying to the crew. In Fig. 3-11 we provide a few of the terms given by various manufacturers to the system that manages this information and organizes it for the flight crew; for reasons of simplicity, we have selected the term *EICAS* (engine indication and crew alerting system) to represent all of them, but we could have selected any of the others, since they all do essentially the same thing. The EICAS does not have authority over all the alerts generated by automated systems on the aircraft, since it is generally limited to major aircraft systems such as engines and powerplant, hydraulic, electrical, pneumatics, and fuel. The FMS is capable of generating hundreds of alerts and messages, as are *decision support systems* (such as onboard weather radar, GPWS, TCAS, GCAS (ground collision avoidance system), and enhanced vision). Most alerting systems have progressed in the past decade from displaying simple failure messages to prioritizing alerts and warnings as well as reducing or inhibiting them during certain critical phases of flight. Increasingly these systems use multiple sensors, large knowledge databases, and sophisticated algorithms to generate flight guidance commands, and in some cases these systems will reconfigure even major aircraft systems such as the powerplants during critical emergencies such as engine failures. The volume of warnings, alerts, and status messages that can be generated by the EICAS, FMS, and decision support systems has reached a level that requires an entirely new approach to organizing and managing this large volume of data for the crew.

One way in which pilots prevent simple failures from becoming complicated is to learn thoroughly and understand the alerting and warning systems of the flight deck: the aural, visual, and haptic[16] cues that direct the crew's attention toward an existing or impending problem with the aircraft's systems or flight path. The way that top performers add to their hardiness for dealing with failure is to thoroughly understand the quick-reference media (electronic or physical checklists and reference handbooks) and how these documents are organized. The methods we have advocated for more than a decade to address both common and uncommon failures have been derived, in part, from our experience in assessing the very best crews and how they manage the wide variety of failures that can occur on the advanced flight deck. In fact, in this specific area, we have specialized in organizing this information for many large organizations, commercial and military, into glass cockpit–compatible *flight crew interfaces* (checklists, quick-reference handbooks, SOPs) to make it easy for crews to find, follow, and execute recovery procedures for system failures in flight. We feel so strongly about the integration of airmanship with information management that from the very first public introduction of Automation Airmanship we have stressed the essential role of *harmonized* flight crew interfaces in developing a truly integrated advanced flight deck.[17]

One of the easiest ways to become complacent as a crewmember on any advanced aircraft is to cede the ground that was once held by the reliable and trusty flight engineer entirely to the automated systems monitors. Recovering that ground can be as simple as knowing the layout of the emergency checklist as well as its logic, organization, and relationship to the aircraft's alerting systems. We have seen it countless times under examination in line checks and simulated emergencies: crews with a detailed knowledge of their flight crew interfaces resolve even the most complex emergencies as though they were all simple, linear systems. Top performers continuously demonstrate one of the best ways to ensure that their own actions in the face of failure are reliable: in addition to knowing their systems, they know how the EICAS logic and other alerting systems work in *cooperation* with the flight crew interfaces. Thus they stubbornly confront the cool and relentless ability of the aircraft with a reliable, steady discipline for dealing with failure that is repeatable over their entire careers and is portable from aircraft type to aircraft type.

Fly First

Equipped with a detailed and reliable knowledge of their own human abilities and limitations, and with a deep understanding of the way their aircraft communicate failures to them, expert glass cockpit pilots have a decided performance edge in the real world, which they know, to paraphrase James R. Chiles, is made up of forces that are still a lot more powerful than they are. However, the principal characteristic of the best performers in our industry, when facing failures in flight, transcends all other knowledge of the phenomenon of in-flight failure, and is present in every meeting with failure that they encounter: they know that to succeed in the face of failure, they must do one thing above all others: *fly first*.

In 2008, the year before what has become a defining year of accidents for our industry with the loss of Air France 447 and Colgan Air 3407, two of the industry's experts summarized in two words what experts know and have been demonstrating for years. Captains Michael Varney and David Owens of Airbus have been questioning traditional training dogma for several years, saying that "Clearly we cannot demonstrate or otherwise teach a crew to deal with every possible risk-inducing situation, but might it be possible to train a crew how to deal with the unexpected in general?" They go on to say,

> Indeed, as aircraft become more complex the number and combination of failures, although less likely, become ever more numerous. With this increased number of potential events, our ability to demonstrate them all becomes impossible in the time available to train in the real world. All we are left with is time to train for the most likely events and time to give general rules, procedures and handling skills to cope with everything else. At the heart of this concept must be the principle of "fly first." How many of history's aviation accidents could have been prevented if the pilots involved did nothing but land the aircraft?[18]

What is so perplexing to experienced pilots when reading the accident summaries, watching computer animated recreations, and participating in briefings of accidents such as those of Air France 447 and Colgan Air 3407 is how both crews overlooked the crucial first step which would have made both accidents mere footnotes in the daily operations of both airlines, instead of the high-profile accidents they have become. Many will view the legacy of these accidents as provoking an overhaul of pilot training, or flight discipline, or even professionalism (all of which have occurred since). We suggest that though automation played a critical role in the flight deck performance of the crews of every accident we have mentioned so far, the fundamental concept of fly first could have altered the flight path and outcome of every one—Air Inter 148, China Air 140, American 965, 85 Victor-Tango, Northwest 188, Tusker 914, Air France 447, Colgan Air 3407, and many others we have not mentioned but that are well known to many aviation professionals. The fact is that other crews have had similar experiences to those crews we have mentioned already, yet applied the principle of fly first successfully. (In Chap. 12 we will look at several incidents similar to that of AF 447.)

Those accidents mentioned where the outcome was not disastrous but rather clearly involved a "fly first" approach to dealing with the problem include US Airways 1549, Emirates 407, and United 232. The crews of each of these flights found themselves confronted with in-flight events that were unique and potentially devastating, whether caused by external environmental factors (a flock of migrating geese), critical flight data input error (a faulty weight and balance entry), or internal aircraft system failure (an uncontained engine failure compromising the entire hydraulic flight control system). Yet each crew managed to control their aircraft by first addressing the integrity of the flight path of the aircraft, and thereby providing the first few critical steps down the pathway of recovery, minimizing aircraft damage and the attendant loss of life. The lessons of these incidents are more or less notorious and provide much to admire in the evaluation of the airmanship of each crew. Yet others, less memorable across the industry, give support to the claim that a "fly first" principle be made fundamental across the vast and increasingly complex environment of twenty-first-century aviation.

Still the Best Buffer against Failure: a Resilient Crew

Skimming just beneath the threshold set by the global media for the coverage of aircraft accidents—perhaps because of a lack of casualties, damage, or dramatic amateur video—is the brief but instructive history of British Airways flight 56 (call sign *Speedbird 56*), which departed O. R. Tambo International Airport in Johannesburg, South Africa, on May 11, 2009, only to return 90 minutes later following the fly first actions of the crew during an unexpected in-flight event that lasted less than 30 seconds.[19]

The Boeing 747-400 was being operated by three pilots and 15 cabin crew and had 265 passengers on board when it began its takeoff roll on runway 03L, just past 1830 local time, for an all-night flight to London's Heathrow Airport. At the controls was the first officer, monitored by the captain, when just prior to Vee-one the number 3 engine thrust reverser REV amber EICAS message illuminated, followed seconds later by the number 2 engine thrust reverser REV EICAS message, at only about 10 knots after Vee-one (which on this night was 150 knots). For experienced pilots familiar with the basic concept of thrust reverse and jet engines, the implications of two thrust reversers deploying after Vee-one (critical engine failure recognition speed) are dire indeed. Short of interviews with the crew, it is difficult to assess what this crew's initial thoughts may have been, to see indications that not one, *but two*, of the aircraft's four engines were not producing takeoff thrust, but were potentially acting in opposition to the operating engines by generating *reverse* thrust, used only in stopping the aircraft when on the runway. The mental model of this situation in the minds of the crew must have included the possibility that the aircraft would most certainly be *in extremis*[20] once airborne, if it in fact became airborne at all.

One of the most practiced emergencies in swept-wing, transport aviation, both civil and military, is the engine failure on takeoff, or the "Vee-One cut." So common is this emergency in training, compared to the relative infrequency of actual engine failures at critical engine failure recognition speed, that some training institutions have reduced its frequency, considering its likelihood so remote that practicing it merely amounts to hardwiring a skill that will likely never be put into practice. However, because of the dire nature of the emergency, it has withstood many assails over the years and remains a staple of every training program, even if it is practiced less frequently than in the past. The crew of *Speedbird 56*, at a moment that many other aviators prepare for over an entire career yet never see, had the advantage of having achieved a level of mastery required by their airline. In fact, they would respond with the "automaticity" of years of training for engine failures on takeoff described by Wickens and Hollands earlier in this chapter (one of the strategies that help crews become *stress-resistant*).

Because of the strict protocols that comprise takeoff procedures (clearly a tightly coupled system, inherently unforgiving, time-dependent, and inflexible), the crew of *Speedbird 56* knew, having passed Vee-one, that they were taking the problem flying along with them, their cabin crew, and their passengers. To attempt to reject the takeoff at such a high speed would most likely have led to substantial aircraft damage, and potential damage to surrounding ground facilities if the aircraft were not to remain on the runway during the abort attempt. What they did not know was that while they were anticipating a powerplant problem, something entirely different was actually occurring with their aircraft. And it had nothing to do with an engine failure or thrust reverser deployment.

Two seconds after the crew received the second thrust reverser REV EICAS alert, the group A leading-edge (LE) flaps retracted automatically just as the aircraft was reaching rotation speed, Vee-R (group A is comprised of the eight innermost LE flaps; six more LE flaps outboard of group A make up the group B LE flaps, which remained extended). The group A leading-edge flaps actually retracted as designed—*whenever the aircraft is on the ground in the "landing phase" of flight*—and the systems detects that the no. 2 and no. 3 thrust reversers are in the deployment stage (this feature would prevent foreign objects and debris caught in the swirl of airflow caused by reverse thrust from damaging these important "high-lift" wing surfaces). Certainly it was never expected that this could occur during takeoff, and most certainly the crew of *Speedbird 56* would not have known of this never-thought-of design shortcoming—however remote and unlikely—during the highly adrenalized and exciting few seconds between the first REV alert and rotation speed. It would be determined later that the indications were caused by "spurious" thrust reverser unlock signals from the no. 2 and no. 3 engines, a possibility that the designers of the "retraction logic" of the B747-400 LE flap system had not identified. With no knowledge that 16 of 28 LE flaps had retracted, the first officer acted just as trained, smoothly rotating the aircraft off of the runway. At 176 knots of indicated airspeed, the huge aircraft became airborne. One second after lifting off of the runway, at an altitude below 10 ft, the stick shaker (an indication of an approaching stall) activated intermittently over a brief time frame of 15 seconds.

> If you want to bring yourself closer to what the crew might have been experiencing during those crucial few seconds, imagine that you have rotated the massive aircraft off the runway, at night, at a high-altitude airport, loaded with enough fuel to travel the length of both Africa and nearly all of Europe; as you do, look at your watch and using the second hand or digital counter, silently count off 15 seconds, while also imagining intermittent stick shaker and the aircraft buffeting of an impending stall.

During those crucial moments the crew would have most likely experienced some level of physiological response to the building stress, including the stimulation of the adrenal glands (which sit like triangular hats on each kidney and are signaled to let loose the powerful chemical cocktail that peak performance requires via the ACTH released by the brain's pituitary gland). Among the chemicals released by the adrenal glands is adrenaline, which serves to increase energy levels and ready the brain-body to handle the threat. (You see, our body is hardwired for survival, and as it would be shown by the pilots of *Speedbird 56*, so are well-trained pilots.)

Immediately after the activation of the aircraft's stall warning system, the captain, as pilot monitoring, began calling out the aircraft height above ground, while the first officer, as pilot flying, controlled the aircraft through the stall warning sequence. At 56 ft and about 182 knots, the group A LE flaps extended automatically, and the aircraft performance returned to normal. Notifying air traffic control of the emergency, the aircraft was climbed to 15,000 ft, where the crew did their best to understand what had occurred, followed procedures to dump fuel, and prepared for the return to Tambo, where they landed uneventfully nearly 90 minutes after they commenced their takeoff roll.

The investigation report is notable in its straightforward evaluation of an automated system that did not perform as anyone might have expected, and likewise in its simple assessment of the crew, stating that "The flying crew should be commended for the professional way that they controlled the aircraft during a critical stage during takeoff and thereby ensured the safety of the 283 occupants on board the aircraft."[21]

No more is said about this crew's expert and exemplary performance, nor has any book been written to chronicle their experience and how they "ruthlessly shed distraction," as has been written of the crew of US Airways 1549's handling of a dual engine failure and their subsequent glide into the Hudson; nor have they been feted in countless interviews and high-profile media events. It is a sparse report for those looking for dramatic content that would point toward a crew caught by surprise by a remote system anomaly that significantly degraded the performance of a near fully loaded jumbo jet, pushing them beyond the peak of their performance abilities into the region of underperformance that has overwhelmed so many other crews. In this incident, more than in any accident or incident we have discussed, is the simple yet profound lesson of how the failure of a complex, tightly coupled system (the thrust reverser operation and LE flap retraction logic) within a similarly tightly coupled larger system (a large aircraft at one of the most critical moments of any flight, at or after critical engine failure recognition speed Vee-one) can be handled by a well-trained crew that clearly chose to fly first. In the split second it took for an automated system to send a spurious signal to another automatic controller which commanded the retraction of critical high-lift surfaces of the 747's huge wing, they found themselves across the border between the domain of "normal" and the domain of "failure." Yet they took the shortest, most effective pathway out of that region by focusing first on flying. They would not know until well after their return what had actually happened.

There are other events like this that have been underreported and otherwise unrecognized. The initial actions of the crew of Qantas Flight 32 over Batam Island, Indonesia, on November 4, 2010,[22] and the heroic flying involved in the safe recovery of a DHL A300-F after it was crippled by a surface-to-air missile in November 2003 in Baghdad, Iraq, are easily researched and tell similar stories. There are more of these "saves" that go largely unnoticed yet have the hallmark of expert Automation Airmanship; and they tell us more about how to respond in the face of failure than the high-profile accidents that make the evening news. We have examined but one, and only at a high level; in keeping with our goal to show how the best crews respond in the face of failure, it stands in stark contrast to the inappropriate actions of the crews of other accidents, whether they are high-profile events whose drama is amplified by the global media or obscure occurrences with little public record.

Our industry may arguably be the best at assigning cause and culpability after the dramatic and tragic failure of the human operator results in disaster. We maintain that although we have much to learn from failure, there is much more to be learned and put into practice by studying success. This requires that those in our industry who have it within their influence to generate new research, to realign safety and training resources to take advantage of the lessons of experienced top performers, and to adopt training and evaluation protocols that allow flight crews to practice, to a high level of proficiency, reliable "fly first" strategies show the leadership to break free of the traditional training ideology that persists in many organizations. The gap that has developed between the human operator and the sophisticated environment of the glass cockpit need not be a problem that requires massive resources and many years to address. To narrow and permanently close this gap across our industry does not require tremendous effort if we collectively shift our focus from institutionalizing the lessons of failure to building into the routines of every pilot and crewmember the habits of success.

Checklist for Success: Positive Flight Path Control

☑ Know the physiology of stress well enough to have a working knowledge of the stress-performance relationship and how the wetware has limits during stressful in-flight events.

☑ During any unexpected in-flight event or emergency, be prepared to "fly first," protecting the aircraft flight path through basic flying skills.

☑ Adopt and practice in-flight strategies for "putting on the brakes," allowing flying, monitoring, and workload to be balanced during emergencies or unexpected in-flight events.

☑ Study and practice with flight crew interfaces (checklists, SOPs, quick-reference handbooks, etc.) until interactions with these resources are smooth, efficient, and consistently errorfree.

☑ Study your aircraft's alerting and backup monitoring systems to a level of proficiency that eliminates ambiguity and unfamiliarity when these systems display information on the flight deck.

Notes and References

1. Captain David Owens and Michael Varney, *Journal for Civil Aviation Training*, 4, 2008, p. 31.
2. ChristopherWickens and Justin Hollands, *Engineering Psychology and Human Performance*, 3d ed., Prentice Hall, Upper Saddle River, N.J., 2000, P. 486.
3. In 1908, psychologists Robert M. Yerkes and John Dillingham Dodson used the term *arousal* for what today researchers commonly refer to as *stress*.
4. Eduardo Salas, Michael A. Rosen, C. Shawn Burke, Gerald F. Goodwin, and Stephen M. Fiore, "The Making of a Dream Team: When Expert Teams Do Best." In *The Cambridge Handbook of Expertise and Expert Performance*, K. Anders Ericsson (ed.), Cambridge University Press, Cambridge, United Kingdom, 2006, p. 443.
5. Joseph LeDoux, *The Emotional Brain*, Simon and Schuster, New York, 1996, p. 241.
6. Bureau d'Enquêtes et d'Analyses, pour la Sécurité de l'Aviation Cevile Interim Report No. 3, Safety Investigation into the Accident on 1 June 2009 to the Airbus A330-203, Flight AF447, Le Bourget Cedex, France, p. 1-4.
7. National Transportation Safety Board, Accident Report, NTSB/AAR-10/01 PB2010-910401, "Loss of Control on Approach," Washington, D.C., pp. 4–5.
8. Wickens and Hollands, op. cit., p. 486.
9. Ibid.
10. Ibid., p. 488.
11. Ibid., p. 489.
12. In a document released by the BEA in June 2011, France's aviation investigative agency, it was stated concerning the final minutes of AF flight 447 that "Throughout the flight, the movements of the elevator and the THS [trimmable horizontal stabilizer] were consistent with the pilot's inputs" and "The engines were working and always responded to the crew's inputs." As for Colgan Air flight 3407, the NTSB's final report found no evidence that the flight controls, engines, and instrumentation had failed during the final 28 seconds of flight or during the events leading up to the accident.
13. "Vee-one," the phonetic term for V_1, as described in the FAA's regulations, is the maximum speed in the takeoff at which the pilot must take the first action (e.g., apply brakes, reduce thrust, deploy speed brakes) to stop the airplane within the accelerate-stop distance. Also V_1 means the minimum speed in the takeoff, following a failure of the critical engine at V_{EF}, at which the pilot can continue the takeoff and achieve the required height above the takeoff surface within the takeoff distance.
14. Charles Perrow, *Normal Accidents*, Princeton University Press, Princeton, N.J., 1999, pp. 75–100.
15. Ibid.
16. Haptics is a tactile feedback technology that takes advantage of a user's sense of touch by applying forces, vibrations, or motions to the operator. Common haptic systems on modern aircraft are stick shakers and stick pushers, which guide a pilot's initial response to an approach to stall.
17. In Chap. 13 we discuss the rigorous process that we have used with a broad range of operators from the military operators (U.S. and others) and civil operators (around the world) to harmonize flight crew interfaces with the technology, culture, and crew configuration of every unique organization and aircraft configuration.
18. Captain David Owens and Michael Varney, *Journal for Civil Aviation Training*, Issue 4, 2008, p. 31.
19. All the facts of this "serious incident" were taken directly from the SACAA's report: "South African Civil Aviation Authority (SACAA) Serious Incident Investigation Report: Accident Incident Investigation Division (AIID)," Ref. CA13/3/2/0717, Final Report. Boeing B747-400 B-GBYA, Group 'A' L/E Flaps Retracted on Takeoff from O.R. Tambo Airport, South Africa, May 11, 2009.
20. *In extremis* is a Latin term used in nautical science to describe a vessel in dire conditions.
21. Ibid., p. 31.
22. The account of Qantas flight 32 is best obtained by accessing the Australian Transportation Safety Board's website at http://www.atsb.gov.au.

CHAPTER **12**

The Ninth Principle: Logic Knowledge

Experts . . . have mental models of [their] equipment. They are not just pressing buttons and receiving messages.

—Gary Klein, 1998[1]

Up to this point we have seen the qualities of expert performers emerge until all that is left to discuss is what we consider to be the last cache of the "secret knowledge of experts." In this discussion of the last principle of Automation Airmanship, we define that singular skill which separates top performers from those whom we could otherwise place in this category based on their ability to apply the first eight principles outlined thus far. It is our experience, and very likely yours as well, that pilots who demonstrate command of this knowledge are those who seemingly "swim away" from the pack with an ease that those left behind can find difficult to emulate. In formulating this final principle, we have identified across the population of top performers that we have observed what we call a *knowledge reserve* that acts as an additional performance multiplier. This knowledge reserve includes information that is not necessarily taught in traditional training, but is obtained through additional study, practice and analysis of pilots' individual experiences in operating the technology itself, and routine study of the experiences of others. Without this additional knowledge, many pilots perform with *adequate* safety margins over most situations; *with it* pilots can experience a level of safety and effectiveness that they can rely on both during routine operations and when situations demand peak performance.

Unreliable Airspeed Redux

As we have seen from the record, 2009 was not a good year for global aviation safety, and for some operators it was catastrophic. Although we had already embarked on writing this book by then, the events of 2009 energized us to complete our compilation of Automation Airmanship principles so that crews everywhere could find in one place the results of decades of experience of expert performance in advanced aircraft. The loss of lives in unnecessary accidents is one very strong motivator for what drives all of us onward toward a "zero accident" future. Yet outside the noise and drama of four major accidents in transport aviation within a span of less than 5 months² from February through June of 2009,² other similar situations were being handled with skill and discipline, informed by the same kind of logic knowledge we introduce in this chapter.

Three incidents investigated in the United States and Australia involving unreliable airspeed, all at altitude, took place between May and October 2009. Each incident involved A330 aircraft similar to that flown by the crew of AF 447, and each incident had strikingly similar circumstances involving weather, high altitude, and multiple autoflight and air data anomalies. Each of the three aircraft was recovered safely, and their crews' valuable reports have since helped operators of the A330 family of advanced widebody jets to implement maintenance and training protocols to enhance safety and reliability. Of these three incidents, the performance of one crew stands out for their actions during a daytime flight from Hong Kong to Tokyo while in cruise flight at 39,000 ft in the vicinity of Kagoshima, Japan.³ While actively engaged in maneuvering their aircraft clear of a convective weather system, the crew of Northwest Airlines flight 8, using the adjustable-tilt, manual and auto modes of their weather radar to gain an accurate depiction of an area of convective activity along their route of flight, entered an area of cirrus clouds (composed of ice crystals). Shortly after entering the clouds, they encountered light turbulence and periods of moderate and intense rain and hail. These conditions alone would capture the full attention of any crew. The alarms and warnings that would occur next would add substantial workload to an already tense flight deck. What happened next and how this crew confronted a serious degradation to their A330's autoflight system help us to illustrate the components of logic knowledge that comprises the heart of this high-level principle of Automation Airmanship.

To get to the level of knowledge embodied in this principle does *not* require that individual pilots become familiar with the complex system schematics, wiring diagrams, maintenance manuals, or complicated process control software that describe in granular technical detail how modern aircraft are built and how they function. In fact, no training program that we know of takes familiarization with the automation to this level; however, we think that more training programs *should* include *enough* information about the logic of the autoflight, flight guidance, and flight management systems to raise the floor of basic understanding to a level higher than that which we see in the current state of glass cockpit initial, transition, and upgrade training. Additionally, those organizations that have the resources to practice and rehearse complicated autoflight failures in full-motion, full-visual simulators should provide their crews with scenarios that allow pilots to refine their skills at handling the aircraft at high altitude, with both normal *and* degraded autoflight and air data systems. Those training programs or pilots that decide not to pursue this knowledge must simply accept that they will be practicing the discipline without the advantage that top-level performers routinely rely on.

We convey two goals in this chapter on the ninth principle: to show in concise terms what the best performers know and to demonstrate how they exercise that knowledge professionally in a way that allows them to continue to build on their expertise, and separate their performance from that of the average and even the above-average crewmember.

Two Kinds of Logic Knowledge

We are applying the term *logic* to not just the *electronic logic* of the automated systems, but also the *procedural logic* of normal, abnormal, and emergency checklists and procedures that determine how the hardware and software are employed in both normal and nonnormal situations. This is not an attempt to simply expound on two similar concepts within the same margins in the interest of saving time and space; it is the realization that, on the advanced flight deck, expert performers know that both types of logic knowledge, practiced together, multiply their performance across virtually every phase of flight. Experienced top performers in any domain not only know how to execute the procedural steps that lead to superior outcomes, but also understand the details of the procedural steps, allowing them to skillfully blend procedural knowledge with system knowledge to perform at a consistently high level.

The Logic of the Automated Systems

The word *logic* can have several meanings across aviation, but for the purposes of this book, we suggest that the definition used in the *Oxford English Dictionary* works very well in our domain and ensures that we do not stray far from the academic interpretation of logic:

> *Logic*: a system or set of principles underlying the arrangements of elements in a computer or electronic device so as to perform a specified task.

This is a concise definition of logic that we can easily apply to the logic that comprises the electronic devices employed by aircraft automation. Let us also recall the definition of *mode* (the one we chose to use in this book) from Chap. 3:

> *Mode*: a collection of system actions associated with different system states.

So, the term *mode logic knowledge* refers to knowledge of the underlying arrangements of the aircraft's automated systems that are responsible for the actions associated with different system states. Victor Riley (whom we have quoted in Chap. 9) says that ". . . modes involve the coordination of avionics 'behaviors' among a collection of devices or functions. In aviation, flight control modes are the most common example of system-level mode."[4] Table 9-1 describes common flight guidance modes that should be familiar to all pilots. Now we are getting close to a practical definition for the flight deck, but so far we only have a variety of sources with varying views on what exactly comprises logic, modes, and mode logic, and all of them are better suited for use by hardware and software designers than for discussion and practice on the flight deck.

To tease a working definition out of all this would not take long, but a simple, elegant, and concise definition has already been provided through the work of three prominent researchers in the field of digital avionics systems: Anjali Joshi, Steven P. Miller, and Mats P. E. Heimdahl. In fact, if you asked most high-performing, expert glass cockpit pilots, they would most likely prefer this simple explanation to any other more scientific, yet complicated definition: *"Mode logic* consists of all the available modes and the rules for transitioning between them."[5] This simple explanation is worthy of committing to memory—and adding to the list of things about your aircraft of which you should have intimate knowledge. Out of this definition we can distill the essential, expert-level knowledge that many high-performing pilots know and practice: all the available modes and the rules for transitioning between them.

In our workshops and presentations on the subject, it is at exactly this point that we gather in the questions and concerns of the attentive, dedicated professionals present and reassert that, yes indeed, the top-performing, most admired and emulated glass cockpit pilots do, in fact, know all the available modes available to them through the various interfaces on the flight deck (to include decision-support systems such as radar, EGPWS, TCAS, HUD, and EVS), and they know the rules for transitioning between them, including "off-normal" and infrequently seen modes. Plain and simple, they do. We know what you are thinking: "Seriously, I have to learn all the available modes and the rules for transitioning between them?" Actually, no, you do not—unless you want to match your performance over time with that of the best pilots in the profession.

And there is more that experienced top performers add to this knowledge that helps them leverage what they know about mode logic through skillful monitoring, situational and mode awareness, and ultimately sound decision making. To transform this knowledge into operational actions, expert glass cockpit pilots know what they are *looking at* and *listening to* through their displays, indicators, and aural alerting systems. Recall that in Chap. 3 we discussed the basic standards for the use of color on the EFIS displays in the cockpit; and we added that this information is supplemented with information specific to individual aircraft that deviates from the standard or is in addition to the basic standards that most manufacturers adhere to; and you guessed it—experts know what every symbol and each color of displayed information represent, in addition to its source. In other words, experts know whether the VSI (vertical speed indicator) information is generated by the aircraft's Air Data system or IRUs (inertial reference units). If IRU data on the primary flight display (PFD) (for example, speed trend, track, and flight path angle) are a specific color, say, lime green, experts know that too. They even know that if the box or arrow around the altitude changes to amber during a level off, it changes color at 150 ft off altitude, and not 500 or 300 ft. For experts who need to make sense of the status of the aircraft flight path by looking at their indications, especially when experiencing casualties to the system that leave some data either incomplete or missing altogether, *one look is all it takes.* In short, colors and symbols count, and so do the source data that generate them. Experts know these relationships, and what the implications of color and symbology mean for modes that are active, armed, preselected, or simply unavailable. Here is what Joshi, Miller, and Heimdahl say about how all this knowledge connects to mode logic (and applies to most glass cockpits regardless of manufacturer):

The mode logic determines which lateral and vertical modes of operation are active and armed at any given time. These in turn determine which flight control laws are active and armed. These are annunciated, or displayed, on the Primary Flight Displays (PFD) along with a graphical depiction of the flight guidance commands generated by the FGS [flight guidance system]. The Primary Flight Display annunciates essential information about the aircraft, such as airspeed, vertical speed, altitude, the horizon, and heading. The active lateral and vertical modes are annunciated at the top of the display.[5]

In pursuing a high level of mode logic knowledge, first know what modes are possible and the rules for transitioning between them and then learn what they look like on the cockpit displays and how they behave when they are active or armed. When you are in training, make a special effort to commit this knowledge to memory; when you are fully qualified, no matter how long you have been "in the seat," practice the use of this knowledge by comparing and updating your mental model of how the automation should be carrying out your instructions to it, and how it is actually doing. Experts do not just sit there, pressing the buttons and receiving messages.

It is important to know, as all experts will attest, that a mode annunciation on the PFD does not always mean that the pilots can expect the exact same behavior every time it is displayed; the aircraft flight control computer will command a different level-off geometry, for example, when the aircraft vertical speed is 10,000 ft/min versus 1000 ft/min, during a descent to a cleared-to altitude. The same applies to the radius-of-turn computations that command bank angle to the flight guidance system during a turn to capture the final approach course of an ILS (or during other lateral navigation operations). Even the most basic FMS function—a turn between two database fixes along a flight plan—has elements that experts need not have detailed knowledge of, as long as they have a *fundamental* appreciation for how the FMC coordinates with the FCC to keep the aircraft on track. (In earlier aircraft designs without an FMS or RNAV, this fundamental of all in-flight maneuvers was left up to the airmanship of the pilots, with some help from the autopilot.) In a technical report for the FAA produced by the Mitre Corporation, researchers Albert A. Herndon, Michael Cramer, and Kevin Sprong describe the fundamentals of a turn between two waypoints, a basic turning path design element or radius-to-fix (RF) leg just like the one depicted in Fig. 12-1.[7]

Paraphrased from the technical report, an RF leg is a constant-radius turn between two database fixes; the inbound and outbound paths are tangent to the arc, and a center fix is also specified. The challenge for the FMC is to compute a path and direct the aircraft, via the flight control computer, to stay on the designed path under constantly changing wind direction and wind speeds requiring varying

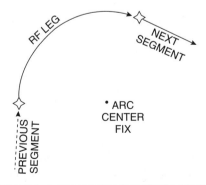

Figure 12-1 Example of a radius-to-fix (RF) leg between two FMS database waypoints.

aircraft bank angles, typically using roll steering. The minimum radius allowed by the design specifications found in FAA orders is determined by a combination of the aircraft category for which the procedure is being designed, a limiting wind value, and a design bank angle control margin of 5°. It is truly staggering the amount of math and science that is engineered into the FMC to make this kind of maneuver possible, where just a few decades ago pilots used lead radials from fixed navaids (sometimes calculated by a navigator) to carry out this operation with minimal deviation from the desired track. Today this is done effortlessly and out of sight of the pilots, and all that we see is a smooth magenta arc with an aircraft symbol centered over it. As we have mentioned, constantly changing environmental conditions such as temperature and wind will have an impact, as will aircraft speed, configuration, and powerplant state. There is no way for you to know how much, but if you understand the basics of the RF leg requirements, say, in unusually high winds, you will understand the limits of the airplane's automated systems to comply with a constraint—and likely lead to actions to stay within your clearance limit, terrain limits, or other constraint. Mode logic enforces the constraints that sequence the tightly synchronized modes which provide effective guidance of the aircraft, and skilled, expert pilots back it up. The algorithms that make this possible are complicated and beyond most pilots; but knowledge of both their existence and the demands placed on them allows pilots to act as better backups, and to more tightly integrate themselves with the technology. This approach puts experts in line to do exactly what the machine cannot—maintain a posture of flexibility and adaptability that the wetware is best suited for. During routine, day-to-day operations, this kind of knowledge will result in fewer track deviations (and potential ATC flight violations) and missed constraints; on a "bad day" with Murphy in the jump seat, this kind of knowledge just might keep you in the game.

Even two FMCs made by the same company but installed on airplanes made by different manufacturers will behave differently from each other because of combinations of the software with the hardware of engines and flight controls. It is impossible, we posit, to expect a crew to mirror the ability of the autoflight systems to make these critical computations. As the United States and other major regions of the world transition to performance-based navigation (PBN) airspace, some amount of mode logic knowledge will become even more vital for pilots. In this environment (elements of which are already in place in the form of tens of thousands of PBN en route, arrival, approach, and departure procedures worldwide), experts learn and apply the *fundamental logic* of mode engagements, mode capture, and mode transitions so that they can act as dependable backups when the automation fails (rarely) or when error or miscalculation by the crew (much more probable) causes a condition that violates those basic parameters.

Flight Control Laws

We have made some vague references to the term *flight control laws* up until now, but it is time to introduce the concept and how knowledge of this area of automation factors into mode logic knowledge for expert glass cockpit pilots. Operational modes, sometimes called *computational laws,* assign flight control laws during flight. These computations are complex mathematical operations that ultimately are manifest in the movement of the aircraft's flight controls to control the aircraft along the flight path commanded by the pilots. Even most test pilots do not need to know the language of these calculations; expert pilots, however, must demonstrate their command over the flight path of the aircraft by understanding the foundations of this family of logic. Since you have come this far in your search for knowledge of mode logic, consider what researchers Joshi, Miller, and Heimdahl have to say about how the automation exerts its capability on controlling the flight path of an aircraft in any lateral or vertical maneuver that the flight crew has commanded through the FMS, FGCP, or other input device to the aircraft's automation:

> Some modes . . . distinguish between capturing and tracking of the target reference or navigation source. Once in the active state, such a mode's flight control law first captures the target by maneuvering the aircraft to align with the navigation source or reference. Once correctly aligned, the mode transitions to the tracking state that holds the aircraft on the target. Both the capture and track states are sub-states of the active state and the mode's flight control law is active in both states.[8]

This description could be for any number of situations desired by the crew: selecting "heading hold" on an intercept heading to join an ILS course which has been armed for capture by the crew; commanding a 500 ft/min descent to a selected altitude; "arming" the LNAV capability of the flight guidance on the runway prior to takeoff in order to comply with an RNAV SID (which would likely involve an RF leg or two as well).

Flight control laws vary by manufacturer, but are nominally considered to dictate how much automatic control is permitted during a respective phase of flight and how the aircraft will respond to inputs by the pilots. Airbus has seven basic laws (normal, ground, flight, flare, alternate, direct, and mechanical). Boeing, on the other hand, favors only three (normal, secondary, and direct). Other manufacturers follow different schemes, and you should know what the designers of the aircraft that you fly specify in their own flight documents and training materials. A sometimes spirited debate surrounds the benefits of each design philosophy, but we will not enter that debate or even favor one or the other; plainly, they all have advantages and limitations. Our viewpoint favors only one system—the *human system*—and that system needs to be tightly integrated with whatever design philosophy it encounters on the flight deck. With any modern, glass cockpit aircraft there are many combinations of modes, submodes, and control laws, many of them interacting with the flight guidance system, speed control/autothrottle system, and installed safety systems. Listing them all here would be futile; but experienced top performers—and those who would emulate their behavior on the way to becoming experts themselves—seek this knowledge as it applies to their own aircraft and make it part of their knowledge reserve.

Procedural Logic

At the risk of alienating the lexicographers at the Oxford University Press, we simply shorten the definition for *logic* given at the beginning of this chapter and substitute *steps* for elements and *procedure* for computer or electronic device. After this, we are left with a very concise definition of procedural logic:

> *Procedural logic*: a system or set of principles underlying the arrangements of steps in a procedure so as to perform a specified task.

Not unlike its distant electronic kin the mode, there are different kinds of procedures, and it is vital that experts understand what they are. It might also surprise a few readers to know that there is an active and prolific field of research that specifically focuses on the design of effective procedures which helps pilots understand what goes into the design of effective procedures and why. A prominent collaboration of researchers from government, academia, and industry continues to produce excellent insight into this crucial area of design: Loukia Loukopoulos, Key Dismukes, and Immanuel Barshi write about their research in *The Multitasking Myth: Handling Complexity in Real-World Operations* (2009) and suggest that there are six main types of procedures:[9]

1. Initial configuration and testing of aircraft systems before flight
2. Engine starting and shutdown
3. Programming of the Flight Management Computer (FMC)
4. Control of aircraft movement through space by manipulation of flight control surfaces and engines
5. Navigation, using onboard electronic systems and signals from external sources

6. Checklists, used in conjunction with the other types of procedures, to ensure that crews do not omit or incorrectly perform items, especially those critical to safety or effective performance

Note that abnormal and emergency procedures are contained in the quick-reference checklists (or electronic checklists on some aircraft) provided by the manufacturer or an organization's own flight department, found in category 6 of the list above. We consider this a special family of procedures that require careful design and training to produce the desired outcome in abnormal and emergency situations. In any case, each of the six types of procedures has a special procedural logic associated with it: whether it is simple, serial, and noncritical or if it is tightly coupled and requires some steps to be conducted in parallel as crews work to contain abnormal and emergency procedures from endangering the aircraft, crew, passengers, and cargo. We have also found scarce efforts by many training departments to equip their crews with this logic knowledge; and we feel it is becoming increasingly important as systems become more complex and safety systems proliferate across the industry. It is not enough to know where to find the right checklist or procedures— knowing how it is organized and how the arrangement of steps contributes to the successful completion of the specified task is the objective. Here are ways that experts learn and master this kind of logic knowledge:

1. Study and practice with flight documents, references, and checklists to develop a sense of how and why certain steps are in the order they are, the vulnerability of procedure to certain types of human error, and the general layout of the graphical information for easy, reliable "navigation" of these documents.

2. Generate frequent feedback on their understanding and execution skills of all types of procedures from check airmen, instructors, and their colleagues.

3. Memorize where critical procedures are located in checklists and quick-reference handbooks (knowing the "geography" of their checklists and other crew interfaces is a specific subskill of this principle).

4. Practice or rehearse complex abnormalities and emergencies such as complicated electrical bus failures, unreliable airspeed/air data emergencies, and FMS/FMC casualties of varying degrees.

5. Use the time they or their organizations invest for training and proficiency to practice complex failures in simulated operational conditions.

In Chap. 14 we discuss in detail our work with organizations and colleagues from diverse backgrounds in developing philosophies, policies, procedures, and practices for highly automated aircraft, and for organizations of all sizes that desire to optimize the relationship between the technology and the operator. As tempting as it is to skip ahead to that chapter, it is important that we stick to the subject and assume for now (naïvely) that every aircraft comes with ideal procedures and practices for all normal, abnormal, and emergency situations that can arise over the course of the flying time of a particular aircraft type. (We will discuss reality with regard to that statement in Chaps. 13 and 14.) Whether ideal or not, the research validates what our experience concludes—that training and evaluation programs for advanced aircraft need to better emphasize procedural logic and knowledge. Again, NASA researchers comment on the topic:

> We must recognize that procedures and training that are based on the assumptions of the ideal operating environment are not helpful, since operators out in the field have to face the complex reality of the operation and the discrepancies between the ideal and the real. . . . Not only must the procedures be designed with a thorough understanding of the complexity of real world operation and the strengths and vulnerabilities of the real operator, but also the training must reflect that understanding. Part of this understanding is recognizing the ways in which the prototypical error-conducive situations manifest themselves in the particulars of the given operation. Another part is recognizing how the performance of individuals and teams is affected by the inherent characteristics of human cognition, organizational factors, and the interaction of these factors with cognitive characteristics.[10]

We have seen the results when logic knowledge is demonstrated at mastery levels on the flight deck. We have seen it combined with other principles of Automation Airmanship and the subskills that make them effective countermeasures to an increasingly hostile operating environment. We can now complete the narrative of Northwest Airlines flight 8 over Kagoshima in the summer of 2009, so eerily similar in both the onset of the emergency and what the indications were to the crew of Air France 447 that one must wonder why there has been so little attention given to this crew's successful actions just three weeks after the fatal crash of AF 447. Our analysis looks at the situation through the lens of *logic knowledge:* mode logic and procedural logic that the crew employed to prevent a second catastrophic outcome in the fashion of a 4.5-minute plunge from altitude.

Soon after their encounter with severe weather at altitude—in spite of their proactive efforts to avoid it—the crew of Northwest flight 8 received a master warning and master caution alert, disengagement of the autopilot and autothrottles, and disengagement of both pilots' flight directors. (*Hand-flying a swept-wing widebody at high altitude requires intense focus and concentration, even when all the indicators are functioning properly.*) Yet in addition to the casualties already mentioned, this crew experienced airspeed fluctuations on both primary airspeed indicators as well as the standby indicator. They experienced a stall warning, noticed that the flight law switched to Alternate Law, along with several other messages related to air data and navigation systems. Although it was daytime, they remained in IMC (instrument meteorological conditions) with no reliable visible horizon.

The FAA's Advisory Circular that provides guidance to industry on flight standards for the design of advanced aircraft electronic flight deck displays contains tables of safety objectives for failure conditions of certain key flight instruments.[11] For airspeed failure conditions in which all primary displays and standby displays are lost, the table lists the hazard classification as *catastrophic*. For just the loss of the primary display, the hazard classification is listed as *major-hazardous* (recall that the investigation reported that the crew of Northwest 8 temporarily lost *both* primary and standby displays). Similarly, display of misleading airspeed information on both primary displays, when coupled with loss of stall warning or loss of overspeed warning also carries with it a hazard classification of *catastrophic*; and for a loss of barometric altitude primary displays, it carries a hazard classification of *major-hazardous*. The same tables describe the required design probability for these malfunctions as *extremely improbable, extremely remote*, or *remote*. In the advisory circular, the FAA has set some high design standards for display technology, in particular these occurrences and those like them, and classified them as having potentially devastating results if they occur in flight. Even with the low probability of these events occurring, in a time span of less than 6 months, similar events occurred in three other A330s operated by three different carriers (TAM, Air France, and JetStar). Fortunately, the crews of Northwest 8 and the TAM and JetStar flights were ready to confront the failures and safely recovered their aircraft.

For Northwest flight 8, pitch and altitude deviations during the first part of the incident (which lasted about 35 seconds) were less than 150 ft, as recorded by the flight data recorder (FDR) and quick access recorder (QAR); flight control laws then returned to normal briefly, and for 10 seconds the crew were able to engage the autoflight and turn farther from the weather. After this brief respite, similar fluctuations returned and lasted for another 3 minutes, during which a climb of about 250 ft was recorded. During the event, following procedures from the quick reference checklist, the crew "controlled the airplane by pitch and power reference" until they were clear of the severe weather, the fluctuations ceased, and they were able to reengage the autopilot and autothrottles, even though the flight law remained in alternate. Clearly the crew combined expert airmanship with expert logic knowledge to contain a potentially hazardous (or worse) condition, including a good measure of *fly first* skill as well. Unfortunately, the lessons of these good outcomes are not as likely to be learned by the vast majority of pilots worldwide, as are high-profile accidents characterized by the tragic and dramatic loss of life.

We often advise that the best time to become familiar with emergencies that involve complex electrical problems, FMC/FMS casualties, flight control malfunctions, and erratic, unreliable, or suspect air data indications (altitude or airspeed in particular) is *at zero ground speed and* 1.0*g*, in other words, before detailed familiarization with the procedure is required in flight, in actual operational conditions. As we said earlier in this chapter, experts know this and take their fates in their own hands when they study these kinds of failures and casualties, and practice or rehearse them in training or during low-workload, long-range cruise when they could be engaged in many other more pleasurable activities.

In this final chapter on the principles of Automation Airmanship, we feel confident in both our observations as instructors, evaluators, and check airmen that we can offer the following list of 10 areas of logic knowledge that experts aspire to acquire over their entire careers, spanning a variety of advanced aircraft, whether they serve as captains, first officers, or augment pilots. After a decade of consideration, study, and research into human factors and technology, we know of no shortcut to obtain this knowledge. It requires self-study, rehearsal, and practice. Wherever possible, this practice should be as part of or in conjunction with initial, transition, upgrade, and continuation training in ground school, briefings, and simulator training and evaluation.

In learning and practicing this knowledge, pilots will be living the ninth and final principle of Automation Airmanship discussed in this book.

So we have come to the end of the description of the nine principles of Automation Airmanship that, as stated in Chap. 1, are designed to not only get you into the game of successful glass cockpit airmanship, but also keep you improving as you progress through your career, experiencing more and more demanding flight operations in an increasingly wide variety of advanced aircraft with the promise of deep and lasting satisfaction in the job. If this text contained everything you need to know to be successful, it would be much longer and encompass many more volumes; much of that additional information is contained in the SOPs, flight handbooks, flight crew operations manuals, and other operational documents provided to pilots for the specific aircraft type that they fly. At a minimum, we think this book sets a standard of professionalism for individuals and organizations, large or small, that is easy for professional pilots to understand and just as easy to demonstrate or observe in all facets of operations.

Of the nine principles, we do not have any favorites, or suggest that any is more valuable or essential than any other. A skilled expert blends them together over the course of every flight, whether it is a long transoceanic crossing of many hours or a short leg between small airports. The principles have been learned and documented over a decade of close observation and close-up interaction with many fine professional aviators, and we are certain that their collective practice will immediately and permanently change the way you approach your interaction with technologically advanced aircraft. We know this because it has deeply affected our own practice of airmanship, and has been embraced by many of those whom we have had the privilege to introduce to these concepts over the past decade.

In the final two chapters of this book we add a dimension to Automation Airmanship that clarifies the high personal standards of expert performance that you can adopt to measure your personal effectiveness in operating glass cockpit aircraft. In addition, we describe the real-world application of these principles and how they have been woven into the standardization and training programs of several diverse organizations, demonstrating the practicality and durability of some of the concepts advocated in the first 12 chapters of this book.

Checklist for Success: Logic Knowledge

☑ Learn the basic laws and phases of flight (or comparable term) that the aircraft is designed around.

☑ Know the basic modes that the aircraft operates in (in each of pitch, roll, and speed) and the individual submodes that make up each primary mode.

☑ Know how modes and submodes can combine and/or be used together to control the aircraft's flight path right now, and allow the pilots to configure the flight guidance for future changes in the flight path (course intercepts, descents, level-offs, etc.).

☑ Memorize altitude level-off logic (during climb, cruise, descent, approach, and go-around). Because this is such a crucial function of automation and accounts for many of the clearance violations involving automated aircraft, it is worthy of its own category.

☑ Know the aircraft's descent logic: descent from cruise, the behavior of the flight guidance system during cruise, and descent to alert height, decision height, or decision altitude (or minimums) on an instrument approach.

☑ Know the autopilot and autothrottle connect and disconnect logic: what steps it takes to smoothly connect the autoflight systems, what conditions cause them to connect or disconnect independent of pilot actions, and how to smoothly disconnect them without adding distraction or flight path deviations.

☑ Learn the indirect mode-change logic, which results from delegating tasks to the automation: what conditions permit changes in flight control laws or phase of flight laws automatically (for example, from cruise to descent, approach to go-around).

☑ Learn the operator authority limits: the boundaries allowed by the design of the autoflight systems which prevent or resist pilot input in order to maintain the aircraft flight path within design maneuvering limits.

☑ Know the feedback mechanisms to the pilots including the salient cues and indicators that display mode states, "armed" modes and mode changes on the cockpit instruments (FMA changes, for example, or the PFD indications of preselected conditions such as airspeeds, headings, and altitudes).

☑ Know how installed "decision support systems" (such as EGPWS, WAGS, TCAS) interact with the flight guidance system: exactly what pilots can expect to see on their PFD for flight guidance during activation of these systems. For example, does the system use flight guidance commands to indicate desired pitch? Or does a successful outcome rely on the pilot ignoring the normal flight guidance commands in favor of evasive action during a terrain, wind shear, or traffic avoidance maneuver?

Notes and References

1. Gary Klein, *Sources of Power: How People Make Decisions*, MIT Press, Cambridge, Mass., 1998, p. 153.

2. Four major accidents from February 2009 to June 2009 include Colgan Air flight 3407 on February 12, Turkish Airways flight 1951 on February 25, FedEx flight 80 on March 23, and Air France flight 447 on June 1.

3. NTSB report DCA09IA064 and ATSB Transport Safety Report, Aviation Occurrence Investigation AO-2009-065 Final, Unreliable airspeed indication 710 km south of Guam 28 October 2009 VH-EBA Airbus A330-202.

4. Victor Riley, "Reducing Mode Errors Through Design," *Commercial Avionics*, March 2005, p. 2.

5. Anjali Joshi, Steven P. Miller, and Mats P. E. Heimdahl, "Mode Confusion Analysis of a Flight Guidance System Using Formal Methods," University of Minnesota, Rockwell Collins and the NASA Aviation Safety program, and the Langley Research Center under Contract NCC-01001, p. 3. *Proceedings of the 22st Digital Avionics Systems Conference (DASC '03)*, Indianapolis, Ind., October 16, 2003, p. 2.

6. Ibid, p. 2.

7. Albert A. Herndon, Michael Cramer, and Kevin Sprong, "Analysis of Advanced Flight Management Systems (FMS), Flight Management Computer (FMC) Field Observations Trials, Radius-to-Fix Path Terminators," The MITRE Corporation's Center for Advanced Aviation System Development, McLean, Va., 2006, pp 1–4.

8. Op Cit, p. 2.

9. Loukia D. Loukopoulos, R. Key Dismukes, and Immanuel Barshi, *The Multitasking Myth: Handling Complexity in Real-World Operations*, Ashgate Publishing Limited, Farnham, Surrey, England, 2009, pp. 22–23.

10. Ibid, p. 132.

11. FAA AC No. 25-11A, *Electronic Flight Deck Displays*, June 21, 2007, pp. 28–30.

Part V

Practical Applications for Automation Airmanship: A Decade of Experience from the Field

There is solid evidence that the highest levels for performance in a given domain are not stable but sometimes continue to increase over historical time as a function of progressively higher and more effective levels of training and practice.

—K. Anders Ericsson[1]

CHAPTER 13

From Average to Excellent: Performance Benchmarks for Twenty-First-Century Flight Operations

In general, the knowledge of top performers is integrated and connected to higher-level principles.

—Geoff Colvin, 2008[2]

When we began our work on the concepts embodied in the *Nine Principles of Automation Airmanship,* it was not enough for us to simply define the elements that comprised each of the nine principles, bring our viewpoints to industry through our contact with fellow aviators in both large and small groups at symposia and workshops, and then move on to the next engagement. We imagined other aviators at organizations of all sizes, operating any variety of advanced aircraft, as wanting what we ourselves would: a precise description of how Automation Airmanship is put into practice through the entire life cycle of an individual flight or mission and, in turn, over the span of an individual career in flying—preferably with examples. We were fortunate to have two factors that weighed heavily in our favor in developing the nine principles and subsequently in writing this book: access to several organizations transitioning from legacy aircraft to the latest generation of advanced aircraft, as well as a team of individuals from a range of diverse organizations with experience in developing training and evaluation programs for advanced aircraft while also being current instructors and evaluators in a variety of advanced aircraft, civil and military. We also knew that we were coming at the problem from an entirely new direction—and would be learning a few things along the way. Our thirst for a concise, representative body of knowledge for our fellow aviators and a broad search of best practices in similar high-risk/high-reliability fields led us to investigate the extensive research into expert performance elsewhere. What we have since learned through studying the elements of the rising discipline of expertise and expert performance was exactly what we needed to shape the principles into an "index of expertise"—contemporary airmanship's own knight's tour—and, ultimately, the performance standards of Automation Airmanship. With this kind of assessment system we could (and later would be able to) assess individuals and organizations and provide both performance and process feedback, as we would help them develop strategies to move smoothly into twenty-first-century aviation.

Limiting our work to simply making bold propositions about how pilots and their crews should change their approach to airmanship as technology changes the way we fly was not an option: we had to be able to make the changes take hold across organizations that we worked with by also updating the traditional "institutions" that have stood for decades and comprised the very core of how organizations employed both advanced aircraft and highly trained crewmembers in accomplishing their mission. In Chap. 14 we discuss some of the more "concrete" institutions as well as the institutional (and cultural) change that we have developed into a rapid and systematic way to update the tools and aids that crews use to interface with their aircraft (everything from normal, abnormal, emergency, and combat checklists to broad organizational orders). However, in this chapter we focus our analysis on how we have helped individual pilots, private flight departments, and military organizations blend the principles of Automation Airmanship into normal, routine flight operations, at the level of personal performance. In an effort to illustrate how this can be successfully accomplished, in this chapter we examine just one component of one organization's successful implementation of the principles of Automation Airmanship across its widespread and diverse operation. For us, stopping our work at the point where we had gathered the key elements of expert glass cockpit airmanship into a neat, concise family of principles would have been mere platitude. Pushing beyond just words and into the cockpits of advanced aircraft with verifiable actions reminds us of a popular saying that is applied to improving individual and team performance: "We don't think ourselves into a new way of acting, we act ourselves into a new way of thinking."[3]

One of the most important values that drove us to broader application was our own personal experience with the wide variety of "—M" programs of the past two decades (CRM, crew resource management; TQM, total quality management; ORM, operational risk management; TEM, threat and error management; etc.). All have made contributions to safety and efficiency, good in both concept and intent, but each with its own complications, and each often difficult to implement at the operational level. We knew that the principles into which we had organized the knowledge of the very best, expert glass cockpit pilots were teachable, observable, repeatable, and scalable; and we knew they could be carried into practice at various levels of effectiveness, similar to the broad concepts of CRM, TQM, ORM, and TEM. We also wanted to ensure that there was a way to make them widely available to pilots lacking access to the training resources that most large organizations rely on to bring new knowledge and new concepts to their pilots and crews.

Our study of exceptional performance (within aviation and other high-risk/high-reliability professions) and how to put it into action brought us into contact with the writings of executive management expert Marcus Buckingham (quoted in Chap. 1). Simply borrowing from Buckingham's observations of what comprises a "controlling insight," not only should we ensure that the nine principles guide action, but also to be truly effective Automation Airmanship should serve as a *multiplier*, across a wide range of situations: ". . . it must serve as the multiplying factor that elevates average to excellent, and it must lead to more precise actions."[4] We were not seeking to create something "additive" to training, standardization, and evaluation programs. It had to be a discipline that could be enacted through training, evaluation, and standardization programs *already in place*. It has never been one of our objectives to replace any of these or other related disciplines, but rather to make them all have greater impact and be more effective by focusing on the qualities of excellent performance of those top-tier pilots who routinely performed complicated tasks on advanced flight decks with unencumbered elegance. After all, even under the closest scrutiny, the nine principles have been practiced in our midst now for many years, just not by all of us, and without the kind of authority that is characteristic of a concise discipline.

The Five Levels of Automation Airmanship

Once we had established the foundational principles, we were faced with defining how we could describe the various levels of expertise as evidenced through actions on the flight deck. Surely, as crewmembers become more experienced and skilled, there would be individuals who put the skills associated with them into practice with a high level of skill and authority, and others who performed at lower levels. In short, we wanted to provide, to those we were working with, a systematic and deliberate approach to recognizing and evaluating Automation Airmanship as it occurred in its natural setting, similar to the way expert performance is evaluated and translated in other high-performance domains, and help them locate their abilities along a continuum of lifelong professional achievement. We needed a system that would support the adoption of these principles according to a measurable standard, as described by K. Anders Ericsson: "Once we conceive of expert performance as mediated by complex integrated systems of representations for the planning, analysis, execution, and monitoring of performance, it becomes clear that its acquisition requires a systematic and deliberate approach."[5]

We needed to let individual aviators, and those who worked with them, know exactly what they were looking for and how to categorize the levels of performance they were observing. We had lots of choices in devising a workable reference for our own "index of expertise." But for ease in implementing a practical performance scale, it was decided that there would be five levels, each defined by how well an individual performed skills that we associated with each of the nine principles.

Performance Standards for Automation Airmanship

Level 1 Performance and understanding of automated systems are significantly below accepted, safe, and efficient levels. Crewmember(s) require constant correction and are unable to effectively manage most automated systems. Interaction with automation is clumsy and slow and negatively impacts the performance of other crewmembers.

Level 2 Performance and understanding of automated systems are mostly correct and appropriate, with occasional deviations caused by inadequate knowledge and understanding of automated systems and/or procedures. Crewmember(s) can be trusted to accomplish basic functions using FMS, FGCP and MCP, and EICAS by rote. Transitions between levels of automation are accomplished with additional time and effort, and they can require assistance from other crewmembers.

Level 3 Performance and understanding of automated systems are safe and efficient, and foundational logic is understood and put into practice using accurate, correct, and standard practices. Interaction with aircraft automated systems and other crewmembers is smooth and accurate, with only rare deviations that are immediately corrected. Transitions between all levels of automated flight are smooth, including manual aircraft control during all normal operations.

Level 4 Performance and understanding of automated systems clearly and convincingly demonstrate above-average knowledge of automated systems and procedures. Errors made by the crewmember(s) are seen before they impact the flight path. The crewmember(s) is/are relied on to instruct and correct others. Transitions between all levels of automated flight are seamless, even during high-workload phases of flight, emergencies, and abnormal situations.

Level 5 Knowledge of automated systems reflects a detailed technical understanding of the underlying architecture of procedures and flight deck technology, and this knowledge is applied to flight deck operations. Additional study and research of the latest, contemporary developments into technology and human performance is undertaken as a routine and is shared with contemporaries, increasing their knowledge and skills. There is flawless interaction with FMS, FGCP, MCP, and EICAS, and decision support systems, through all input pathways during all phases of flight. Manual aircraft control is as smooth as that guided by automation and flight guidance.

The first thing that comes to mind when most individuals are exposed to this set of standards is that level 5 Automation Airmanship seems to be largely unattainable by most pilots, and that the average performer, doing his or her job as any manager or supervisor would expect, would likely be categorized as performing at level 3. Even though level 3 performance is intended to mark "the center of the bell curve" it has hallmarks of what many experts demonstrate routinely. Admittedly, not everybody is a top-performing expert, and as we have seen, even the most experienced pilots (as measured by flight hours alone) can perform at level 1, for a variety of reasons which may include lack of proficiency, fatigue, chronic stress, or some other cognitive or physical impairment. As with any measure of this nature, a lively and spirited debate followed choices made in composing this index, and the conclusions from any further debate would likewise not be unanimous.

Our experience with these benchmarks over the past eight years has proved to be reliable and requires little change; it has been modified and adapted to several unique settings with success, and it continues to serve as a sound measure of performance. In 2005–2006, with the help of our colleague Dr. Tony Kern (whose "Historical Airmanship Model" has withstood the test of decades of practical use) it was adapted for use as an evaluation tool by U.S. Coast Guard C-130J crews as they prepared a vanguard of pilots and crew to transition to one of the most advanced military transports of this century, the Lockheed-Martin C-130J family of aircraft (a dramatic and near-complete transition of this venerable classic airframe into one of the most advanced, capable airlifters in modern aviation).

In 2008 it was used as part of an evaluation instrument to assess the automation "readiness" of Royal Canadian Air Force crews as Canada embarked on one of the most ambitious, comprehensive fleet upgrades of any Western air force in modern history. In formulating these standards we discarded many additional descriptions of both excellent and poor performance in favor of brevity; we acknowledge that there are many more ways to describe each level of performance, especially as more is learned about the exact nature of expertise in the coming decades.

Early on we were encouraged by observations and findings from high-risk/high-reliability domains (energy, medicine, and transportation including aviation), professions, or pastimes with a clear hierarchy of performance abilities (music, sports), and domains typified by complexity and risk (finance, military operations, etc.). Not everybody in these domains is an expert, and even some experts do not manage to perform at the highest levels. Our observations, both formal and informal, reveal the same basic pattern, and we feel strongly that to become truly expert at operating advanced aircraft, newcomers and veterans both must be willing to push beyond the plateau that defines "normal excellence" and pursue the more difficult path toward lifelong achievement. Again, the best of the research into the expert performance suggest that

> After months of experience, [novices] typically attain an acceptable level of proficiency, and with longer experience, often years, they are able to work as independent professionals. At that time most professionals reach a stable, average level of performance, and then they maintain this pedestrian level for the rest of their careers. In contrast, some continue to improve and eventually reach the highest levels of professional mastery. . . . Further improvement depends on deliberate efforts to change particular aspects of performance.[6]

One of the most useful models we have encountered in describing what this looks like is the performance-experience model of K. Anders Ericsson, shown in Fig. 13-1.

Ericsson's performance-experience model helps to explain the difference between experts and nonexperts and is easily adapted to twenty-first-century aviation and Automation Airmanship. Ericsson goes on to describe the qualitative difference between experts who improve their performance over time and those who achieve and maintain a more "pedestrian" level of performance over a career:

> The goal for everyday activities is to reach as rapidly as possible a satisfactory level that is stable and "autonomous." After individuals pass through the "cognitive" and "associative" phases, they can generate their performance virtually

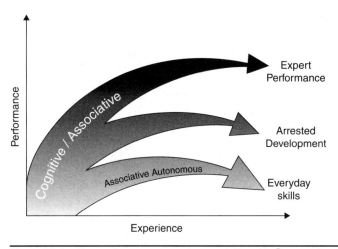

Figure 13-1 The performance-experience relationship.[7]

automatically with a minimal amount of effort (see the gray/white plateau at the bottom of the graph) [see Fig. 13-1]. In contrast, expert performers counteract automaticity by developing increasingly complex mental representations to attain higher levels of control of their performance and will therefore remain within the "cognitive" and "associative" phases. Some experts will at some point in their career give up their commitment to seeking excellence and thus terminate regular engagement in deliberate practice to further improve performance, which results in premature automation of their performance.[8]

In essence, it takes additional effort and work not just to reach, but also to maintain level 5 or even level 4 Automation Airmanship. It is consistent with our observations, given that we have been exposed to aviators of all experience levels, in both civil and military settings: Being in the top tier of all performers requires more than just what is required by most training programs and organizational standards of "acceptable" performance. As we stated earlier, not everybody will perform at the level of the best experts; part of the challenge is to set the median performance standard higher than it has traditionally been by implementing some of the most common performance multipliers of the best among us. In short, the benchmark performance characterized by levels 3, 4, and 5 is higher than many organizations have maintained in previous decades, and asks that the vast majority of crewmembers begin to adopt the performance standards that the best experts follow. Because of what is known by both formal and informal inquiry into high performance, we are unapologetic about what we consider to define the "mean benchmark" in practicing twenty-first-century Automation Airmanship. We would not be so confident of its practicality, if we had not described in detail the specific areas that these standards can be applied to (across the nine principles), making it possible for nonexperts to understand what it takes to perform at a higher level.

Although we believe we have simplified much of the perceived complexity of the modern glass cockpit in earlier chapters, on a personal level we can attest to the simple fact that although the road to the highest level of performance is clearer, there are no shortcuts to *maintaining* a high-level performance over time. There are elements of performance levels 3, 4, and 5 that require individuals to go beyond what they may be used to in maintaining more traditional levels of airmanship in less complex aircraft, or in less-demanding environments. Organizations that expect to keep their crews on the forefront of safety and operational effectiveness have an obligation to ensure that opportunities for deliberate practice, study, and debate of the latest knowledge related to both technical and nontechnical skills which allows experts to obtain high levels of performance are offered to the entire organization. The increasingly demanding global aviation environment, the rapid adoption of highly automated aircraft worldwide, concerns over the global pilot shortage for the first part of the century, and the prospect of continuous upgrades and improvements to existing aircraft systems are just a few of the factors that, in our view, require individuals and organizations alike to ensure that they take steps to reach beyond traditionally accepted levels of expertise and performance. Simple economics, the competitive marketplace, an increasingly error-intolerant culture, and the demands for ever-increasing levels of efficiency are just a few of the pressures that promise to be unrelenting in sorting out those who adopt high personal and professional standards from those who opt for the status quo. In his 2008 book *Talent Is Overrated: What Really Separates World-Class Performers from Everybody Else,* Geoff Colvin reinforces the distinction between "high performers" and the vast majority of their contemporaries, whether in business, aviation, medicine, performance art, sports, or hundreds of other high-reliability professions, and what that means to business: "For virtually every company, the scarce resource today is human ability. That is why companies are under unprecedented pressure to make sure that every employee is as highly developed as possible."[9]

One organization with which we have worked for over 6 years has taken unprecedented steps to integrate the nine principles into the way it instructs and evaluates human performance across one of the most diverse and advanced aircraft fleets ever assembled. The impact that this will have on the development of its next generation of aircrew will surely help the organization fend off many of the contemporary pressures placed on it.

The Royal Canadian Air Force: Raising the Human Performance Bar

As we stated in the opening lines of this chapter, our commitment to seeing the broad principles of Automation Airmanship put into universal practice goes beyond a few industry talks and some occasional contact with aviators in the course of our day-to-day jobs within the air transport industry. You just cannot take a new idea to your peers and hope they appreciate the high concepts and begin to adopt them, putting the ideas thoughtfully into practice as they themselves advance the practice of contemporary airmanship. And you certainly cannot take the podium in front of hundreds of your colleagues from across the industry without science, consultation with research, and a deep understanding of the mechanics of the modern training delivery chain that supports twenty-first-century flight operations. Additionally, the concepts have to be able to be adapted to the unique culture of the individual organization with ease. Of all the myths we have encountered in a decade of working with our colleagues across the industry, one of the most powerful and compelling ones is embodied in the hundreds of commercial offerings of commercial off-the-shelf (COTS) and military off-the-shelf (MOTS) aircraft and aircraft systems.[10] We are firm in our judgment that no advanced aircraft can serve even seemingly similar organizations with the same training, flight crew procedures, flight deck practices, and interfaces. In just one example among others, after extensive experience working with Lockheed-Martin C-130J operators in the USAF, USCG, USMC, and RCAF, we can safely say that the COTS/MOTS approach to selling advanced aircraft is little more than marketing in many respects. As Henry Petrosky, the prolific professor of engineering at Duke University, has said, "Things work because they work in a particular configuration, at a particular scale and in a particular culture."[11] In Chap. 14 we will share our experience with the "hard" science of creating twenty-first-century, practical tools for the advanced flight deck from traditional "legacy" institutions; we finish this chapter with a discussion of what is often considered the "soft" side of flight deck performance, but increasingly holds greater influence over outcomes in contemporary flight operations.

In 2007–2008 we worked with our colleagues at Convergent Performance and Esterline-CMC to assess the readiness of the Royal Canadian Air Force (RCAF) for an unrivaled transition from an aging aircraft fleet to one of the most diverse and advanced combinations of aviation resources of any Western air force, including fixed-wing, rotary-wing, tactical, and strategic transport and strike/fighter as well as unmanned aerial vehicles (UAV). Adapting the principles of Automation Airmanship to an assessment tool that could be applied across a variety of aircraft and operational settings, we sought to identify the strengths, vulnerabilities, and gaps within the RCAF as they adopted steps to establish the service as a global leader in operating advanced aircraft. After an aggressive 12-month investigation that took us across the entire nation of Canada to interact with virtually every aircraft community, the results were reported to command leadership in the "Automation Analysis Report."[12] Among its many findings and recommendations, it examined the shortcomings of the Human Performance in Military Aviation (HPMA) program, administered by a separate *"flight"* under the RCAF Central Flying School, both in its then-current state as the RCAF's human factors training program and as a potential multiplier for the service's transition from legacy to advanced aircraft. Subsequent to the report's findings, the commander of the First Canadian Air Division ordered that the HPMA program be updated and infused with the principles of Automation Airmanship. Thus the RCAF would become the first modern air force to observe, evaluate, and report on crewmember proficiency and expertise on the modern flight deck, from ab initio training to an individual's last check ride in any one of Canada's modern aircraft. As one senior commander would say later, "In fact, HPMA and automation airmanship become the same measures and standards in a well-developed automation culture."[13]

We decided at the outset that Automation Airmanship should be able to fit into most rigorous programs without adding training time, infrastructure, or a sustaining presence in addition to what most organizations already have in place for both technical flight training and human factors instruction and evaluation. And that is exactly what we found in the RCAF's HPMA flight: a cadre of aviators well grounded in human factors and with an institutional role already embedded in the training system. All we needed to do was to adapt the nine principles of Automation Airmanship to fit the culture, scale, and configuration of what was already in place. There was no need to "throw out" the existing principles upon which the HPMA program had been built (including such CRM staples as decision making, team performance, threat and error management, etc.); leadership within the RCAF Central Flying School determined that not only these traditional CRM mainstays would be updated and retained, but also performance measures and standards would be developed for skills associated with Automation Airmanship. In the end, the program would use existing infrastructure to deliver up-to-date, leading-edge human factors training and evaluation for every RCAF aviator.

Similar to other Western air forces, the RCAF relies on a core cadre of experienced, accomplished senior pilots and aircrew to administer training and evaluation and to ensure standardization in every fleet. Known as the Standardization and Evaluation Team (SET), the SET from each aircraft community meets regularly (at least annually and sometimes twice a year) in Winnipeg at the headquarters of the 1st Canadian Air Division to share knowledge and participate in programs designed to enhance training, evaluation, and safety across the RCAF. For several years leading up to the development of the performance measures for the HPMA program, alongside our counterparts from Esterline CMC's Human Factors Engineering group in Ottawa, we had already been working with this group of resident experts, developing and implementing a new family of Air Orders and Policies for the entire RCAF. Literally thousands of hours of effort were spent collaborating with some of the most professionally motivated and knowledgeable women and men in the Royal Canadian Air Force, all of them responsible for the successful adoption of new technology as new aircraft replaced aging models, and some aging aircraft received comprehensive technology upgrades. Using this already proven model of success, we teamed up with this group in 2010 and 2011 to develop from existing HPMA and Automation Airmanship what the commanders wanted: practical guidance and tools for helping pilots and aircrew across the entire RCAF to get their increasingly technically demanding jobs done better, and with an additional margin of safety. Over months of working groups with the SETs and collaboration with the HPMA team, a concise family of "behavioral markers" was developed over 18 skill areas totaling close to 400 individual, specific crewmember actions defining crewmember performance from preflight to postflight in advanced aircraft of all types. With this "library" of behavioral markers, the HPMA Flight in Winnipeg would be poised to make immediate and lasting changes to how crews are trained and evaluated in operating complex aircraft in demanding domestic operations as well as combat operations abroad.

Behavioral Markers for the Twenty-First-Century Flight Deck

The decision to embed Automation Airmanship principles into the Royal Canadian Air Force's human factors program through developing detailed, specific behavioral markers is just one way for one organization to adapt its existing infrastructure to multiply the effectiveness of its crews as Canada replaces or upgrades much of its operational aircraft fleet in less than a decade.[14]

But just what does a "behavioral marker" do, and how does it accomplish its purpose? In short, behavioral markers describe specific, observable crewmember behaviors (*not* attitudes or personality traits), with a clear definition that makes it easy both to learn and to observe in practice. A good behavioral marker, by virtually any accepted authority with knowledge of them, has a causal relationship to a desired performance outcome in the organization's operational environment (say, maintaining situational and mode awareness during a multiple-resource search and rescue mission). In the case of the RCAF human factors program, the goal was to deepen the meaning of each human factors skill area (including the principles of Automation Airmanship) by defining the specific actions that each individual aircraft SET for their respective community expected of its entire crew as they employed their aircraft in an operational setting. In short, "this is how we describe what we expect our crews to be able to do." Just a few of the payoffs include data relating to performance, training assessment, safety management, performance trends, research applications, and of course a robust system for providing feedback for individuals, groups, teams, and the organization. The multiplier effect of using a robust library of behavioral markers as the backbone of a rigorous evaluation system rapidly promotes the organizational goal of developing and maintaining a culture of excellence with respect to all facets of airmanship. And the expectations for performance are transparent, for instructors, evaluators, and trainees alike.

In Table 13-1 we provide a representative "slice" taken from the RCAF's behavioral markers project that shows how specific and detailed the SETs sought to define the performance objectives for their crews in just one of the 18 HPMA skill areas (6 existing skill areas plus 12 more which are inclusive of the nine principles of Automation Airmanship). Applying rigorous standards for developing behavioral markers while working closely with the "experts" in the RCAF in a controlled and monitored process resulted in performance baselines for flight deck and mission compartment crewmembers that are rigorous and measurable; can be tailored specifically to individual aircraft configurations, crew complement, and other operational factors; and are sensitive to levels of performance as observed by both the evaluator and the trainee, with the same transparency. Like many of the other activities that we have been associated with in putting Automation Airmanship into broad use in organizations, we found that there are no shortcuts to long hours of guided activities and detailed documentation with experts in their natural setting, if the job is to be done properly. The list of 24 behavioral markers, "automation situational and mode awareness," in Table 13-1—developed in collaboration with the RCAF Human Factors in Military Aviation (HPMA) Flight, Convergent Performance, and Esterline-CMC Human Factors Engineering team—is an example of the depth that an organization can go to when it wants to ensure that its members will be able to reach very high levels of performance in operating the most modern of aircraft in the most demanding conditions of modern combat and peacetime operations.

Combining this type of inventory with carefully considered levels of performance is not just a powerful recipe for performance evaluation (and all its applications), but to paraphrase Marcus Buckingham, such a tool can serve as the multiplying factor that elevates average to excellent and leads to more precise actions. We turn now to one of the most important ways in which an organization

1. Strictly adheres to information display SOPs
2. Uses appropriate level of automation for each phase of flight
3. Uses displays (CNI-MU/HUD/HDD) that reflect crew intentions
4. Selects display scale that enhances mental model
5. Sets appropriate scale on displays
6. Keeps displays uncluttered and relevant to phase of flight
7. Knows meaning and significance of all symbology on primary instruments
8. Ensures that displays are up to date and uncluttered
9. Demonstrates accurate mental model consistent with automation
10. Seeks updates to mental model to maintain ASMA
11. Demonstrates cross-check procedures
12. Avoids automation fixation
13. Anticipates mode changes
14. Announces mode changes
15. Visually and verbally confirms mode changes
16. Calls out FMA changes
17. Ensures that mission compartment communicates specifics of changes to flight path guidance
18. Promotes understanding between monitoring and the level of automation selected
19. Ensures that"automation surprises" are rare
20. Adheres to limits
21. Announces surprises immediately
22. Changes level of automation in use to maintain safe, controlled flight
23. Debriefs automation surprise events as soon as possible after resolution
24. Uses written references to augment knowledge during abnormal or degraded operations

TABLE **13-1**   The RCAF HPMA Behavioral Markers for Automation Situational and Mode Awareness (ASMA)[15]

can begin to extract valuable information about flight deck performance simply by overturning one of the most entrenched practices in all of flight training.

Performance Evaluation for the Twenty-First-Century Flight Deck

Careful examination of Table 13-1 reveals a particular quality of our approach to airmanship for the modern cockpit (and the contemporary mission compartment, wherever appropriate) and a kind of evaluation that is different from what most of us are accustomed to. One of the subtexts of this book, from the beginning, has been that our profession—and others that share some of its nuances—is such that we stand to learn more about expert performance by examining successful strategies than by studying failures. Unfortunately, the reporting system (regulatory and voluntary, across the organization and individually) is tilted toward accidents, incidents, near-accidents, and near-incidents and on substandard performance during both operations and training. With few exceptions, for most aviators it is difficult to bring to mind very many notorious occasions where true expert performance resulted in an outcome that the rest of us can model our actions on; instead we are constantly bombarded with assessments of failure. As we stated in Chap. 12, we stand to learn more about top performance from the actions of the crew of Northwest flight 8 over Kagoshima than from the inactions of crews that fail when facing similar situations, no matter how intense the government, media, and public scrutiny. We find this to be the same in training programs that we have come in contact with as well.

If you are an instructor, evaluator, check airman, standards pilot or crewmember, training system designer or regulator, chances are that your evaluation system focuses predominantly on reporting performance and gathering data on substandard performance and failure. Any aviator who has ever taken a check ride knows this all too well. Recording outstanding performance is most often an afterthought, provided as "additional comments" and rarely takes the form of capturing top performance and categorizing it as it occurs, in the proper context, complete with the nuance that makes such performance unique. In the succinct words of Gary Klein, "We put too much emphasis on reducing errors and not enough on building expertise."[16] As we have learned and observed, the most widely respected and effective trainers and evaluators are skilled at both— correcting deviations and guiding learners to high levels of performance and expertise. Again, in 2009 Gary Klein commented that

> Because we know more than we can tell, trainers gravitate toward describing the explicit parts of the work. These are easier to describe. We find it easier to give feedback about errors than about skillful actions or about improvements in mental models.... In short, when we try to improve performance we usually emphasize explicit knowledge more than tacit knowledge. In giving feedback, we tend to focus on specific procedures. These tendencies aren't problematic in well-ordered situations. However, they are inadequate for improving performance in complex situations.[17]

If we can leverage the science of expertise and expert performance to increasingly and accurately describe the "tacit" knowledge of experts, why can we not influence how performance is reported, data are collected, and the lessons of top experts fed back to the entire profession? Klein's work in the last decade supports an increasingly growing enthusiasm for training instructors and evaluators to notice the context of performance as well as the outcome. This approach can result in a richer account of performance than by simply noticing how often mistakes are made, balancing a fear of mistakes with the pursuit of expertise.[18] We think that in an era of increasing change and competitiveness driving more and more technology into the hands of aviation professionals, this is one of the great opportunities of our age to positively influence both safety and performance. If we are going to seize the opportunity, then we should design training that encourages the demonstration of both tacit and explicit knowledge, train instructors to value developing expertise as much as error avoidance, and provide models for expert action to learners across the entire experience spectrum.

Reaching beyond Normal Excellence

In this book we have discussed only a few of the thousands of examples of profoundly expert flight crew actions that have proved to be critical in determining the successful outcome of an in-flight emergency. We have made some bold assertions about how these experts compile their skills and experience to handle these crises, and now we suggest how organizations and individuals can pursue the kind of training that can promote this kind of top-level performance. It is not the kind of training many of us have become familiar with, even in the age of line-oriented flight training (LOFT), advanced qualification programs (AQPs), and similar situation-based training. In 2005 we became familiar with NASA's Emergency and Abnormal Situations (EAS) project which was on the cutting edge of research into different types of emergencies, from the most mundane to the rare situation for which little guidance is provided to the crew in the form of checklists, procedures, and prior training. Its goal was to ". . . develop guidance for procedure and checklist development and certification, training, crew coordination, and situation management, drawing on knowledge of the operational environment, human performance limitations and capabilities, and cognitive vulnerabilities in real-world emergency and abnormal situations."[19] We were inspired by the challenge of the project's broad objectives, and we were stirred to action in our own research into training and interface design by some of its early findings, specifically with regard to automation: knowing which level of automation to use and other ways in which automation can pose a threat to the successful outcome of an emergency or abnormal situation. Unfortunately the project was canceled prematurely; but its limited reports and publications are consistent with what other researchers have found to be important factors in the successful handling of complex, uncommon, and unexpected in-flight situations. Not surprisingly, these findings conclude that practicing complex problems in the simulator results in more successful outcomes in flight, even among expert crews. In 2004 Earl McKinney and Kevin Davis reported on this observation in their research, which includes these recommendations:

> . . . Pilots may need to be trained to explicitly consider higher levels of the cognitive map. For example, in training at higher levels, pilots might be given aircraft malfunctions that are quite random and perhaps distinctly unlikely to occur. This type of training may enhance the ability of experienced pilots to identify cues in their decision process that lead to less effective decision making during the actions selection phase.[20]

The close relationship between practice and outcomes at the highest level of individual and team performance is foundational to the science of expertise; and never in history has our industry had more capable, reliable, and realistic flight simulation than now. Harnessing this technology as a multiplier of expert performance seems only logical; not doing so seems negligent.

If we had the power to bring full funding to research projects such as NASA's EAS project and all the others that might promise to contribute to the acceleration of our understanding of the very best performers, we surely would. But we live in the real world, one fraught with risk and competition, increasingly difficult business economics, and countless other challenges for our industry and others like it. In the face of these factors, we are aware that the most practical way to see Automation Airmanship take hold across the industry is through existing institutions that are already efficient and reliable at providing training and evaluation. Our goal is simply to push the boundaries of what has become entrenched practice, bordering on tradition, within the training delivery chain. In doing so, we believe we will see the day when every glass cockpit aviator can break into the open and perform at similar high levels as those that he or she looks up to and desires to emulate. To borrow once more from one of our age's most prolific researchers and authors on the subject of expert performance, K. Anders Ericsson: "At the highest levels of expert performance, the drive for improvement will always involve search and experimentation at the threshold of understanding, even for the masters dedicated to redefining the meaning of excellence in their fields."[21]

In the last chapter we transition from our experience with evaluation and training of Automation Airmanship in real-world, practical settings to the more concrete and tactile tools found on the contemporary flight deck. We will demonstrate methods that blend the nine principles of Automation Airmanship into real resources that flight crews can rely on to support the highest level of performance on the flight deck. In a similar way, we will demonstrate how we have developed methods for the leaders of organizations of any size to multiply the capability of the flight deck crew as they employ advanced aircraft to increase the organization's broader goals of safety, efficiency, and operational effectiveness.

Notes and References

1. K. Anders Ericsson, "The influence of experience and deliberate practice on the development of superior expert performance." In *The Cambridge Handbook of Expertise and Expert Performance*. Cambridge University Press: Cambridge, UK, p. 690.
2. Geoff Colvin, *Talent is Overrated: What Really Separates World-Class Performers from Everybody Else*, Portfolio Books: New York, 2008, p. 106.
3. Larry Bossidy and Ram Charan with Charles Burck, *Execution: The Discipline of Getting Things Done*, Crown Business, New York, 2002, p. 89.
4. Marcus Buckingham, *The One Thing You Need to Know*, Simon and Schuster, New York, 2005, p. 16.
5. K. Anders Ericsson, Neil Charness, Paul J. Feltovich, and Robert R. Hoffman, "The Influence of Experience and Deliberate Practice on the Development of Superior Expert Performance," *The Cambridge Handbook of Expertise and Expert Performance*, Cambridge University Press, Cambridge, United Kingdom, 2006, p. 698.
6. Ibid., p. 683.
7. Ibid., p. 685.
8. Ibid.
9. Colvin, op. cit., p. 18.
10. One of our counterparts at Esterline CMC's Human Factors Engineering team in Ottawa, Bob Kobierski, originated the idea that "COTS/MOTS only works for the first customer." We could not agree more.
11. Henry Petroski, "Look First to Failure," *Harvard Business Review*, 82(10), October 2004.
12. R. D. Kobierski and C. Stickney, "Automation Analysis Report," Air Force Automation Policy and Planning Development Project, Winnipeg, Canada, September 29, 2008. Prepared for DND under PWGSC Contract No. W8485-0-XKCF/01/BQ.
13. Lieutenant-Colonel Colin Keiver, CD in "Automation Airmanship: Optimizing Aircrew Performance in a Modern Air Force," *The Canadian Air Force Journal*, 2(2): 10, Spring 2009.
14. Recent and planned acquisition plans call for Boeing C-17s, Lockheed Martin C130Js, Sikorsky *Cyclones*, Augusta *Cormorants*, advanced fighters, in addition to "future air vehicles" to replace the DeHaviland *Buffalo* for search and rescue. Existing aircraft receiving comprehensive upgrades to cockpit technology include the *Aurora*, a Lockheed ASW aircraft based on the generations-old and successful P3.
15. Used by permission of the Royal Canadian Air Force. © Her Majesty the Queen in Right of Canada
16. Gary Klein, *Streetlights and Shadows: Searching for the Keys to Adaptive Decision Making*, MIT Press, Cambridge, Mass., 2009, p. 13.
17. Ibid., pp. 46–47.
18. Ibid., pp. 111–112.
19. Barbara K. Burian, Immanuel Barshi, and Key Dismukes, "The Challenge of Aviation Emergency and Abnormal Situations," NASA/TM-2005-213462, Ames Research Center, Moffett Field, Calif., June 2005, p. 11.

20. Earl McKinney and K. J. Davis, "Effects of Deliberate Practice on Crisis Decision Performance," *Human Factors: The Journal of the Human Factors and Ergonomics Society*, 2003. 45(436): 444.
21. K. Anders Ericsson, K. Anders, Neil Charness, Paul J. Feltovich, and Robert R. Hoffman, "The Influence of Experience and Deliberate Practice on the Development of Superior Expert Performance," *The Cambridge Handbook of Expertise and Expert Performance*, Cambridge University Press, Cambridge, United Kingdom, 2006, p. 700.

CHAPTER **14**

Automation Airmanship and Operations: Making Sense of the Technology

What is missing are principles, rules and guidelines defining the relationship between the technology and the humans who must operate it.
—John Lauber, 1992[1]

For over a decade we have been engaged with flying organizations of various sizes, from entire air forces to individual private flight departments, in an effort to help them bring order to the complex environment they find themselves working in with the addition of advanced aircraft to their fleets. In fact, although we have spent many thousands of hours in front of and alongside fellow aviators, encouraging the adoption of the nine principles of Automation Airmanship, we have spent many thousands of hours *more* engaged in making sense of the complex flight deck at the very nexus of human and machine. It is, after all, the "philosophies, policies, procedures and practices"[2] that describe the operational working relationship between the flight crew and the aircraft. These activities have included creating the very broad organizational philosophies contained in foundational guidance such as air orders, policy and standards manuals, aircraft flight manuals, flight crew checklists, and quick reference handbooks. We have been designing every aspect of these operational interfaces, right down to specific procedures such as programming a search pattern in the FMS of a rescue aircraft or flying a precision instrument approach in one of the latest commercial cockpits. We strove, alongside new users, to define the relationship between the crew and the aircraft in a way that both brings out the kind of expert performance described in this book and provides space inside of policies, manuals, procedures, checklists, and other "interfaces" for Automation Airmanship to rise up from individual crew performance and carefully designed flight crew interfaces.

For over a decade, our efforts have been characterized by a range of processes: from those used in a large research project in cognitive science and expert performance, to human factors engineering and ergonomic design, document management, flight test, instructional systems design, and even study of the subtle meaning of words describing technical information in different languages. The necessity to adapt these and other processes into its own discipline is the result of the gap created when aircraft and avionics engineers, designers, and manufacturers produce hardware and software combinations that users must then somehow adopt in minimum time to execute safe, efficient, and economical flight operations in their own operational and cultural environments. In short, *we have been providing a description of the relationship between the crew and the technology, and the myriad details that make that relationship work, independent of any particular manufacturer's philosophy, inside the unique environment in which the crew operates.*

The previous 13 chapters have described some aspects of the multidisciplinary approach we have employed to help describe the contemporary relationship between modern aircraft and the crews who operate them. In this final chapter we introduce several other important concepts and some crucial research which has contributed to the synthesis of what we consider to be one of the most foundational yet often overlooked aspects of modern flight deck design: *the detailed description of the operational transactions between the human operators and the equipment they control.*

The narrative that we use to illustrate this discussion comes from our own experience in the field, working alongside engineers, test pilots, integration teams, technical writers, and most of all, dedicated flight crews whose jobs ultimately involve operating complex aircraft in the most demanding aviation environment of theater warfare. This narrative does not describe the drama of flight deck actions that sometimes can lead to either disaster or heroic airmanship, but a *process* that we remain quietly dedicated to executing and improving through a deliberate and systematic discipline. It is what we consider to be the finishing touch on the aircraft acquisition process, and it has proved to be a true multiplier of safety, efficiency, and operational effectiveness. The combination of these approaches with invaluable findings in the research community has made it possible to bring Automation Airmanship beyond just nine principles or even hundreds of detailed, descriptive behavioral markers demonstrated by expert crews, and into the cockpits and mission compartments of some of the world's most complex and capable modern aircraft.

The Push for Procedures

Make no mistake: we do not think that it is a small leap from adopting a principled discipline of airmanship, as we have suggested, to embedding action steps into procedures and checklists. In fact, it was our experience in devising flight crew interfaces that led to our first introspective efforts to isolate specific actions of highly successful crewmembers that *could not* be embedded into procedures for the advanced flight deck. It has been a very powerful combination for making gains in this unique area of human performance: harmonizing the qualities of the highest-performing crews with the wide variety of technology found on today's modern flight decks, civil and military, fixed wing and rotary wing, old aircraft and new. We think that designing procedures without an insight into the wetware's role in executing them is no more sensible than training flight crews without providing any insight into the capabilities and limitations of the hardware and software. In his 2009 book *Streetlights and Shadows*, Gary Klein sums up the compelling urge to turn good ideas about how individuals should act into procedures:

> The process of transforming skills into procedures is irresistible. All we have to do is break a complex task down into steps and provide some tips about when to start and finish each step. Then we hand the procedures out so that workers can perform this task even without years of practice. Much of our progress in fields such as training, coaching, and safety management comes from this strategy.[3]

A careful examination of Klein's observation in context gives this insight a "tongue in cheek" insinuation, and as we will describe in the rest of this chapter, it is never as simple as the above citation suggests. The terrific urge to proceduralize all that we know in the complex, tightly coupled world of modern aviation is powerfully resisted by the constant need for flexibility and adaptation created by countless outside factors and the fact that even the best procedures are sometimes violated—intentionally and unintentionally—by well-trained crewmembers. Finding the right balance can be more art than science, but it is a process that we are constantly striving to perfect. In fact, much of Klein's work has been devoted to isolating the nuance contained in individual performance, commonly called *tacit knowledge*. Procedures designers have labored for many years trying to proceduralize tacit knowledge, sometimes also referred to as the *knowledge of experts*.

The pressure to use procedures to address safety and efficiency concerns often comes from outside operations, whether from regulators, command leadership, the safety team, or elements of the training chain. From their perch above and outside the day-to-day operation, procedures make it easy for oversight entities and/or command leadership to document an organization's determination to fix a problem, whether it is chronic or acute. The reason is, according to Klein, that procedures can do some pretty powerful things for an organization, among them that "[They] also help us evaluate performance. We can see if someone knows the procedures, and is following them."[4] It is on good authority that an organization's leadership often turns to procedures when a problem in the operation surfaces that threatens safety margins, which have a direct link to the business model of the organization. Degani and Wiener clearly state that "flight deck procedures are the backbone of cockpit operations,"[5] and Klein writes that "Aviation illustrates the ideal arrangement: skilled decision makers living in harmony with procedures."[6] Both views are universally held among aviation professionals and often admired by other high-risk/high-reliability industries outside of aviation. It is just plain hard not to stop at these conclusions since so much of our historical success in arresting accident rates is rooted in devising procedures that turn back quiescent hazards such as distraction, time pressures, and fast-moving situations (and countless more threats to safety) that require well-trained, reflexive operator actions to overcome. Even if the best way to address a hazard to the operation ends up including a new procedure, it is essential both to push past the urge to quickly solve a problem with a procedure or checklist and to consider the nuances of the situation and the role that tacit knowledge (or the lack of it) possessed by the individual factors can play in the outcome. Any of the researchers we have referenced so far would likely insist as well.

In Chap. 12 on the ninth principle, logic knowledge, we introduced three researchers (R. Key Dismukes, Loukia D. Loukopolos, and Immanuel Barshi) who have recently shared their research findings and operational experience in procedures design in their book *The Multitasking Myth*. One outcome of their research represents a key insight into how those outside the procedures design process view the family of interfaces that comprise a crewmember's primary reference for his or her transactions with the aircraft (policy guidance, operating manuals, and normal and abnormal/emergency checklists). From outside the process it *does* seem that in fact aviators are living in a utopian world of "skilled decision makers in harmony with procedures" (Klein) where procedures, as described by Dismukes et al., ". . . implicitly portray cockpit work as having three central characteristics. It is linear, predictable, and under the moment to moment control of the cockpit crew."[7] For all practical pilots and aircrew there is no need to mount a convincing argument that the real world presents a more dystopian view, where countless factors work against such imagined predictability. In fact, Dismukes, Loukopoulos, and Barshi have identified four prototypical ways in which this ideal is upset, troubled, or, in research parlance, "perturbed."[8]

1. Interruptions and distractions;
2. Tasks that cannot be executed in their normal, practiced sequence;
3. Unanticipated new task demands arise; and
4. Multiple tasks that must be performed concurrently.

Their work provides valuable insight into the design process that cannot be overlooked and must be accounted for in the way that aircraft and crewmembers are prepared to operate in the real world, where procedures must be balanced with a robust discipline of airmanship.

There often seem to be two camps in the process to isolate and overcome a problem in which procedure and individual expertise both factor, an argument sometimes known as "human as hazard" versus "human as hero" that has been forwarded and discussed by notable researchers such as Jim Reason, Gary Klein, and others. Our experience in working directly with expert crewmembers in the operational setting where they work every day—complete with complex technology and increasingly complicated crew integration—has reliably shown that there is a close relationship between the two that requires a disciplined process in order to create the appropriate balance. Although there seems to be an unending supply of competing ideas to evaluate the combination of technology and the human operator, the formula we have been able to apply repeatedly with good results for the past decade can be accurately summed up by Gary Klein: ". . . the human-as-hazard model and human-as-hero model are both justified, and we must find ways to balance them. . . . We should find ways to blend them, expanding the role of expertise when procedures and analysis aren't sufficient."[9]

Similarly, Dismukes et al., in their book, support this notion as a result of their own research and practical field experience:

> Simply understanding the relevant equipment and how it is to be used is not enough. Operating equipment safely and efficiently also requires a thorough understanding of the operator, of the environment in which the operations take place, and of the operations themselves.[10]

To find the right mix of analysis, recommendations from the research community, manufacturer recommendations, individual operator constraints, crew composition, and a host of other factors makes this, as we can most certainly attest to, a most interesting endeavor, and one that requires careful integration of the end-user in the process.

In our experience to improve past practices in aviation, we have again found inspiration from credible, widely respected sources *outside* of aviation.

The Pull of Design Thinking

In recent years another rising discipline has helped us to bring our methods of interface design into sharper focus, and has allowed us to get closer to those we have assisted in transforming a complicated and often clumsy cockpit environment into an entirely different dynamic where crewmembers experience early on in their training glimpses of the unencumbered elegance of experts. In a few pages we will offer examples of the discipline we have spent over a decade refining. But before we can be sure that it will be clearly understood, we need to share some of the concepts embodied in what has become known in business and industry in recent years as "design thinking." We first read about the practicality of design thinking in the June 2008 *Harvard Business Review*, and we immediately found a place for the foundational concepts of this approach to product design and business practices in our own collaborative work with users of a wide variety of aircraft. In short, *design thinking* is ". . . a methodology that imbues the full spectrum of innovation activities with a human-centered design ethos."[11] Some of the buzzwords that business and industry were adapting from the design community readily applied to our processes of making sense of the human-machine relationship. We easily added to our existing vernacular phrases such as "beyond features, functions and performance," "scrutinizing user behavior," "breakthrough technology," "multi-disciplinary teams" and "concept generation." Blending these concepts with the imperatives of safety and proper system operations, organizational culture, and operational constraints helped us to generate an output that more completely took into account the end-user. It is, after all, the success of the end-user in operating complex aircraft that the entire design team should be focused on.

To illustrate this concept, we can evaluate two of the many field experiences we have had in the past decade, one of our earliest and one of the most recent. In 2003 we became involved in a collaborative effort with the U.S. Marine Corps as they faced the rapid integration of their first KC-130J aircraft into the existing Marine tanker-transport fleet (which in 2003 was largely comprised of aging "legacy" versions of the C-130). At the time when the Marines were adopting the KC-130J, they were also involved in two theaters of warfare, making the successful integration that much more challenging and important. The "J" version of the decades-old C-130 manufactured by Lockheed-Martin represented a significant diversion from the traditional "steam-gauge" C-130 whose many versions had dominated tactical airlift for nearly half a century: an airplane whose outward appearance seemed nearly identical to its older cousins, the KC-130J has little else in common with its legacy predecessor once an aviator climbs the entryway and steps aboard the aircraft. One of our counterparts at VMGR-252, the Marines' first squadron to fly the KC-130J operationally, was (then) Major Colin Keiver, a Royal Canadian Air Force (RCAF) pilot on exchange with the U.S. Marine Corps who would later go on to be the commanding officer of the first RCAF C-130J Squadron in Trenton, Ontario, Canada, in 2010. After thousands of hours of flying legacy C-130s in the Canadian Air Force and with the U.S. Marines, (now) Lieutenant Colonel Keiver sums up the user experience of the first advanced airlifters delivered to the Marines a decade ago in this way:

> What Lockheed-Martin should have done after building the first C-130J is change its name to anything *but* the C-130J—since its design was so radically different from every previous version of the C-130, changing the fundamental way the pilots and crews approached flying the aircraft.[12]

Just two years earlier we had worked on a similar project with the U.S. Air Force as they began fielding their first C-130Js with the Air National Guard. It was during this project that we came to know many of the pilots and aircrew who would not only successfully bring that aircraft into service, but also over a decade later would be among the first North American users of the C-27J.[13] Lieutenant Colonel Joe Brophy, a veteran of both projects who has flown the C-130J and the C-27J in combat, made this observation recently, when reflecting on his earlier work in designing procedures for the C-130J:

> We assumed the checklists and procedures provided by the OEM were thoughtfully developed to reflect the best practices of the day. We assumed they were refined through flight test and, with the aid of the company's human factors engineers, optimized for the aircraft's military mission and crew complement. These assumptions of course, could not have been more wrong. In fact, the checklists and procedures we received were only designed for the company's test pilots and engineers to carry out test flights and safely ferry aircraft to customers—nothing more.[14]

In our fieldwork with flying organizations around the world, we have found similar viewpoints held by many other end-users of today's most advanced aircraft. The result of the gap between what users need and what manufacturers provide can be catastrophic unless it is addressed early in the adoption phase of the technology. Again, Lieutenant Colonel Brophy, reflecting on the Air National Guard's struggles with the new technology, commented that

> We all sensed a lack of standardization and, particularly among the pilots, a sense that something was missing—we were

expending too much energy and were too "distracted" while performing normal checklist procedures. This gave us concern about our ability to effectively manage complex abnormal situations as well as perform our high-workload wartime mission. This collective sense—strengthened by a comment from one of our senior instructor pilots with "glass" background that he felt he "still didn't know what I was doing" after nearly 100 hours on type—led us to take action.[15]

Although it may seem contrary to traditional methods of aviation procedures development, principles used in the design process are a natural fit for integrating an advanced aircraft into a new or existing organization's aircraft fleet. After all, according to one of the most respected voices in the field of design, Klaus Krippendorff,[16] "design is making sense of things." And one of the most important steps that often must be taken as new technology is introduced is to have it explained—*or made sense of*—including all its nuances and unique attributes, so that both the organization and the operators can make sense of its role in operations. In an industry where new innovations in hardware and software constantly evolve to ever-higher levels of reliability and performance, we find ourselves working very hard to keep pace with the way new technology impacts the experience of the crews who operate it. In the parlance of the design community, these users are called *interpreters*, and they are a vital part of the process of designing compatible, durable, harmonized flight deck procedures and practices.

Between the authors we have shared a common observation that has a lot of truth to it, even if we use the phrase so often that by now it has become cliché within our design team and among those with whom we have collaborative relationships: *If you really want to know how the [new equipment, upgraded system, or other innovation] is going be used operationally, give it to the line operators for a few months—let them expose its practicality and hidden value (or not).* The front-line operators are the interpreters: pilots, mission crew, maintenance personnel, cabin crew, test and evaluation crews, and instructors. In his book *Design Driven Innovation*, Roberto Verganti describes interpreters as ". . . those individuals who give meaning to things; meaning being the emotional, psychological and sociocultural reasons behind products, beyond features and functions and performance."[17] Engaging these individuals systematically in a deliberate process of design not only results in a finely tuned family of crew interfaces, it speeds the development process and has the additional benefit of creating an informed core of "innovation multipliers" who in turn can speed the adoption of both the new technology and new ways to interact with it.

Our discussion of interpreters here is not to suggest the obvious: that these individuals are an important part of the design process and they should have some say in it. That concept has been worked over by business and industry for years, including the aerospace industry, and is nothing new. It is much deeper than that, going even beyond the critical step in any acquisition of putting the new technology into service. We have found that the success of any of the interface design projects we have been involved in has been a close examination of the *meanings* that interpreters extract from their experience and impart on the operational setting created by the introduction of the new technology, regardless of whether that new technology is something as "simple" as the addition of EFB (electronic flight bag) or HUD/ EFVS (heads-up display with enhanced flight vision system) or a completely new aircraft or cockpit upgrade. What the Marines of VMGR-252 experienced in the early part of this century with the advent of the KC-130J was completely different from their experience with all previous versions of the aircraft, as Lieutenant Colonel Keiver elaborated on earlier. Before we could understand how to shape the cockpit experience (through redesigning flight crew interfaces such as normal and emergency/abnormal checklists) in a dramatically different aircraft in terms of technology, capability, performance, and even crew complement, we spent many hours flying with their crews in many of the mission environments where they were operating the aircraft.[18] Not watching and evaluating the experience of these crews close up would have added long hours in revisions down the road. We use interactions like this, as well as others, to "get close" to interpreters, as explained by Verganti:

> The process of design-driven innovation . . . entails getting close to interpreters. It leverages their ability to understand and influence how people could give meaning to things. This process . . . consists of three actions. *Listening* (interacting with interpreters), *interpreting* (assessing the knowledge it gains by interacting with interpreters and then recombining and integrating this knowledge with proprietary insights, technologies and assets), [and] *addressing* the design discourse and diffusing the vision to interpreters.[19]

Using a disciplined, standardized approach that results in opportunities to listen, interpret, and address the experience of individual operators doing cockpit work in their unique operational environment allows us to design interfaces that "tap the full value of the technology."[20]

Tim Brown, President and CEO of IDEO, takes design thinking to even deeper levels by adapting the broader principles into practices that industry has adopted to improve processes in many seemingly unrelated domains. We found their applicability in aviation to be particularly useful, enabling us to further refine steps in our process that already took some of this important knowledge into account.

The hard science of human factors engineering (HFE) is highly adept at task analysis and is capable of producing very informative functional flow diagrams that describe essential cockpit and crew activities, often resulting in the flight crew interfaces provided to users by the manufacturer. We have used this methodology successfully in partnership with the HFE team at Esterline-CMC during a partnership we collaboratively called *totally integrated human factors engineering*. The objective was to streamline the development process in the interest of both time and budget pressure, but had the added benefit of quickly turning engineering findings into practical, human-centered interfaces. This process resulted in much of the foundational material we would eventually use to shape the flight crew interfaces for the HC-130J currently employed by the U.S. Coast Guard. Tim Brown gives some insight to this approach in his book *Change by Design*:

> It's possible to spend days, weeks, or months conducting research . . . but at the end of it all we will have little more than stacks of field notes, videotapes, and photographs unless we can connect with the people we are observing at a fundamental level. We call this "empathy," and it is perhaps the most important distinction between academic thinking and design thinking. We are not trying to generate new knowledge, test a theory, or validate a scientific hypothesis—that's the work of our university colleagues and an indispensable part of our shared intellectual landscape. The mission of design thinking is to translate observations into insights and insights into products and services that will improve lives.[21]

Brown describes the design process with a particular emphasis on the experience of the user and how this can actually factor in improving the outcome of the user's experience with technology. Extending this to the advanced cockpit, we begin to think of improved outcomes as those that not only "improve lives" but that also *save* lives: widening safety margins, increasing efficiency, and providing opportunities for crews to bring the principles of Automation Airmanship to the surface, naturally and without effort. Again, Tim Brown comments that

> . . . we are not only carrying out a function, but having an experience. That function can be compromised if the experience attending it is not designed with the same mindfulness a good engineer brings to a product or an architect to a building . . . Experiences should be designed and engineered with the same attention to detail as a German car or a Swiss watch.[22]

Combining "design thinking" concepts with contemporary research findings and other traditional processes creates a powerful combination capable of truly changing the relationship of the user and the technology.

The Foundations of Good Interface Design

The process that we follow today to create flight crew interfaces that reflect the capability of any aircraft; the size, configuration, and qualifications of the crew; and the wide variety of missions for which the aircraft is to be employed is clearly not the same as when we took on our first challenge in this domain over a decade ago. But back at the beginning of the century (and lucky for us), our efforts were preceded by a unique research team at NASA whose work was ahead of its time and, in our view, widely underused by industry even today. Asaf Degani and Earl Wiener's NASA Contractor Report titled *On the Design of Flight-Deck Procedures* established bedrock principles of interface design, and with their "four P's" (philosophy, policies, procedures, practices) approach to aligning all four of these concepts through careful interface design they blazed the first trail to what many now refer to as *harmonized, compatible procedures* for the modern flight deck.[23] Their report was our first and most reliable reference as we struggled to help organizations make sense of the leap in technology represented in the latest aircraft as they replaced legacy fleets around the world; the four P's became part of our lexicon, and remains so today.

As we have already alluded, the first few organizations we worked with were among the vanguard of North American customers to acquire Lockheed-Martin's fully updated, reengined, and automated airlifter, the C-130J. The U.S. Air Force first acquired the aircraft in 1999, assigning them to the Air National Guard initially, adopting them later for the active duty Air Force. The Air National Guard was followed by the U.S. Marine Corps, who operate a tanker-transport version of the aircraft (the KC-130J) and the U.S. Coast Guard, who operate a search and rescue variant of the same basic airlifter (the HC-130J). By now this highly successful tactical airlifter has been adopted in large numbers by the U.S. military and many other air forces around the world. The lessons we learned in these first few years took us well beyond the four P's, forcing our procedures designers and technical writers to adapt to the needs of each organization's individual operating environment, crew composition, and operating culture before we could shape solutions that would work at the operational level of each unique organization. Although we had completely overhauled normal and abnormal/emergency procedures for one user, we found that the needs of each organization were so widely different in some areas that we needed to go back to the earliest stages in the design process to accommodate the special circumstances of each specific operator. One of the greatest lessons we learned through this process is concisely summarized in the words of Duke University professor Henry Petroski, author of several books on failure and design, who summarized why things work at all: "Things work because they work in a particular configuration, at a particular scale and in a particular culture."[24] Even the guidance we received from NASA researchers did not fully explain how to accommodate differences in operational culture, broader organizational philosophies, and, at the technical level, how individual procedures could be impacted by the different software versions of the flight mission computer (FMC) for each user. We found ourselves going beyond simple procedures design and into technical domains that we had not initially imagined.

One of the toughest hurdles to clear in designing procedures for advanced aircraft is the comparative rapid change that occurs in aircraft capability through new software versions for flight and mission computers. Even small changes in software can have a direct impact on specific procedures and practices that are accounted for in flight operations manuals, standard operating procedures, and normal and emergency/abnormal checklists. The addition to an established cockpit design of equipment such as the electronic flight bag, heads-up displays, enhanced flight vision systems, and other new technology has even greater impact. Keeping track of it all through the interface design process has the potential to overwhelm any team of procedures designers. At this critical crossroads we adopted concepts and principles, normally used in flight test, to ensure proper system function, and quality control steps governed by International Organization for Standardization (ISO) rules[25] as safeguards against unexpected and unwanted consequences from aligning procedures and practices on the flight deck and mission compartment. Using information technology to track individual procedures in effect creates a "living" record of an individual aircraft family's procedures and practices through its entire life cycle. Integrating these processes left us with a repeatable, scalable, economical, and efficient process that we could tailor to virtually any user of advanced aircraft to help organize the complexity of the modern flight deck and mission system. Although we felt we were accomplishing a great deal just by using traditional tools available from within the industry (flight test protocols, ISO practices, reports and research findings from NASA and other agents), it was important to us to redefine what it actually takes to get a new operator of advanced aircraft into a position of greater safety and operational effectiveness through greater integration of its crews with the technology. In essence, the process can be graphically explained as shown in Fig. 14-1.

The simplified procedures design process illustrated in Fig. 14-1 shows the basic relationship between key steps in the process, and the emphasis on the "codesign" approach favored by Tim Brown and IDEO that creates the "us with them" dynamic of the project.

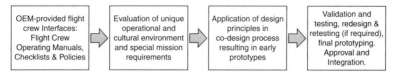

| OEM-provided flight crew Interfaces: Flight Crew Operating Manuals, Checklists & Policies | Evaluation of unique operational and cultural environment and special mission requirements | Application of design principles in co-design process resulting in early prototypes | Validation and testing, redesign & retesting (if required), final prototyping. Approval and Integration. |

FIGURE **14-1**   The interface design process.

Our codesign team normally includes the project's technical writers and project manager, experienced glass cockpit pilots of similar aircraft types with similar missions, new users on the vanguard of the transition team, OEM designers and engineers, veterans from other industry backgrounds, and trained facilitators able to bring experience, research, and knowledge of automation to the codesign team.

The Role of a Prototype Interface

Experience with a variety of end-users of advanced aircraft has shown us the importance of developing an early prototype, one that (to paraphrase Tim Brown) *may not work flawlessly, but that instead teaches us something about our objectives, our process, and ourselves.* The final stage prior to approval and promulgation is one that is particularly important: not only do flight test and validation ensure compatibility between the crew and the hardware, but also, more importantly, rigorous flight test ensures that all procedures are checked against aircraft performance and certification standards.

From August 2010 to August 2012 this process received its greatest test when we were involved in a unique industry collaboration with Global Military Aerospace Systems, Alenia Aeronautica, and L3, the prime contractor for the U.S. military's Joint Cargo Aircraft, the C-27J. It would be the first time that we would overhaul an entire family of flight crew interfaces, completely redesigning them and deploying them to a combat theater at the same time that crews were being trained to operate the aircraft, flight test was being concluded, and new aircraft deliveries were underway to multiple operational units. In an interview with the authors following a combat tour flying the C-27J in Afghanistan in 2012, Lieutenant Colonel Greg McCleary of the 135th Airlift Squadron of the Maryland Air National Guard summed up the codesign process as it was experienced by the end-user in this way:

> The revision process for the C-27J included significant integration of the end-user. With the revision time constraints, the end-users' abundance of prior tactical airlift experience and glass-cockpit experience, in both the civilian and military operations arena, allowed the revision team members to tailor the normal/

emergency checklists to the unique environment and operations required ... Flight test, validation, and working with the original checklists while writing the aircrew training syllabus for the simulator enabled the team to produce excellent revised checklists. We demonstrated this improvement while completing the aircrew simulator training by using the revised checklists and noting the vast differences in ease of usability. Noteworthy is the fact that these revisions started only about 5 months before the simulator training began, and the revisions were complete for the normal/emergency checklists about 2 months into the 5 months of simulator training.[26]

The added pressure and urgency of an ongoing war during the entire design process had an impact on the activity schedule of the design process, but rigorous process and test standards kept this pressure from having an impact on the ultimate output, as evidenced by the experience of the Air National Guard combat crews who flew the C-27J in the "last tactical mile" of contemporary warfare, supplying front-line troops with vital supplies and mobility. Again, Lieutenant Colonel McCleary commented that

The C-27J revised procedures worked and flowed very well during the sustained combat operations. Even so, there were times when normal checklists required interruption or switching to another checklist due to unique outside factors and the unpredictability of combat operations on the daily flight schedule. These situations required more than procedures, they required tacit knowledge to provide flexibility that cannot be "proceduralized." For instance, the ERO [Engines-Running Offload] checklist was written to prepare the aircraft for takeoff after an ERO. However, changes in the flight schedule or changes to the pick-up or drop-off points for passengers and cargo would necessitate other operations or checklists prior to the takeoff. Aircrew would have to use their knowledge of the situation to ensure all applicable items were accomplished in unique situations.[27]

Procedures harmonized with the technology, combined with the reliability and performance capability of a modern, advanced cockpit aircraft such as the C-27J, is a powerful combination. The resiliency of the process and execution of each key step consistent with both the most durable traditional procedures development tools and the research and innovations from outside of aviation made this one of the most successful projects of its kind, and gives the U.S. Air Force the first truly integrated full family of glass cockpit, tactical transport flight crew interfaces, custom-fit to their unique combat requirements.

No matter how economically, rapidly, and efficiently any design team can create a family of compatible flight crew interfaces, ultimate success rests on a few bedrock principles. These come from a variety of disciplines and are represented in Fig. 14-1. They are indispensable if flight crew interfaces are to achieve the same level of reliability and durability as the aircraft hardware and software during both normal and abnormal situations. We have seen the process from start to finish with a variety of organizations and aircraft for over a decade. Even so, the evaluation of this approach to optimizing the relationship between the human operator and the machine is best described by a veteran of not only the design process, but also combat flying in multiple war zones flying two different kinds of aircraft. A final word from Lieutenant Colonel Joe Brophy:

> The irony with the work of developing a technical interface for flight crews is that—for normal procedures—the success of the effort can be judged to some extent by how little it is used. That is to say, well-designed interfaces connect the operator with the equipment in such a way as to minimize dependence on the interface itself. The beauty of this work, when it is done right, is that line instructors teach techniques which are consistent with checklist procedures that are closely aligned with company policies which are a subset of the organization's overall philosophies—all of which is powerful for students.[28]

Pioneers of modern procedures design, Asaf Degani and Earl Wiener summarize the experience with an indispensable, historical perspective:

> We believe that there is no "royal road" to procedure development. There is no such thing as an optimal set of procedures. No manager will ever be able to "open up the box," install the device, and install "good" procedures along with it. Nor do we anticipate that any computer technology can make this easier. Pilots are trained to fly by procedures. Aircraft are built to operate by procedures. Government regulations are based on procedures. It is a long, tedious, costly, exhausting process. We do not know of any shortcuts.[29]

In the fast-paced world of modern aerospace where new technology is introduced, improved upon, and sometimes mandated for adoption by regulators, the pressure to reduce costs and shorten adoption time lines will only increase with time. All of us who are interested in increasing safety margins through improvement of the experience of the human operator need to continuously seek innovative ways to ensure that shortened time lines and cost reduction do not overcome the gold standards of safety and reliability. Where procedures cannot be designed to encourage expert performance that is both safe and reliable, tacit knowledge must be trained so that users will be able to maintain adequate safety margins and flexibility simultaneously. Discipline in both areas, in our experience and in our opinion, will make the greatest difference in ensuring safe, effective, and economical operations regardless of the overall organization or unique capabilities of any modern aircraft.

What we have outlined in this chapter on developing effective flight crew interfaces is not the *only* way, but it is *one* way; and it is a method that works in a fast-paced world. It is not the optimum way, since for us the "optimum" way is always over the horizon. But we do not have to be obsessed with the solutions that might exist over the horizon while we work to solve the challenges that exist for today's flight crews. We expect that the coming decades will provide even more challenges that will require all of us to adapt, as we all seek the optimization of the human-machine interface—where the ultimate goal is to extract the full value from advanced technology for both the operators and those who benefit from it.

Notes and References

1. E. H. Phillips, "Pilots, Human Factors Specialists Urge Better Man-Machine Cockpit Interface," *Aviation Week & Space Technology*, March 23, 1992, pp. 67–68.
2. Also known as the "four P's" (philosophy, policies, procedures, and practices), a concept introduced by Asaf Degani and Earl L. Wiener in their report, "On the Design of Flight-Deck Procedures," NASA Contractor Report 177642, NASA Ames Research Center, Moffett Field, Calif., June 1994.
3. Gary Klein, *Streetlights and Shadows: Searching for the Keys to Adaptive Decision Making*, MIT Press, Cambridge, Mass., 2009, p. 15.
4. Ibid., p. 29.
5. Degani and Wiener, op. cit., p. 53.
6. Klein, op. cit., p. 29.
7. Key Dismukes, Loukia Loukopoulos, and Immanuel Barshi, *The Multitasking Myth: Handling Complexity in Real-World Operations*, Ashgate Studies in Human Factors for Flight Operations, Farnham, Surrey, England, 2009, p. 22.
8. Ibid., p. 106.
9. Klein, op. cit., pp. 106–107.
10. Dismukes et al., op. cit., p. 108.
11. Tim Brown, *Design Thinking*, in the Harvard Business Review. Harvard Business School Publishing Corp., Watertown, Mass., June 2008.
12. Lieutenant Colonel Colin Keiver, Commanding Officer 436 Squadron, First Operational C130J squadron in the RCAF, in an interview in June 2012.
13. The Alenia C-27J Spartan is a medium-sized military transport aircraft, an advanced derivative of Alenia Aeronautica's G222 (C-27A Spartan in U.S. service), with the engines and systems of the Lockheed-Martin C-130J Super Hercules. The aircraft was selected as the Joint Cargo Aircraft (JCA) for the U.S. military.
14. Lieutenant Colonel Joe Brophy, in an interview with the authors in July 2012.
15. Ibid.
16. Klaus Krippendorff is the Gregory Bateson Professor for Cybernetics, Language, and Culture at the Annenberg School for Communication, University of Pennsylvania, Philadelphia.

17. Roberto Verganti, *Design Driven Innovation: Changing the Rules of Competition by Radically Innovating What Things Mean,* Harvard Business Press, Boston, 2009, pp. 3–6.

18. Prior to our work with the Marines of VMGR-252, we had been involved in other interface design projects for the C-130J family of aircraft, including similar efforts with the USAF Air National Guard and U.S. Coast Guard. In every case, our preliminary task analysis resulted in different interface design for each individual user.

19. Verganti, op. cit., p. 13.

20. Ibid., p. 65.

21. Tim Brown. *Change by Design: How Design Thinking Transforms Organizations and Inspires Innovation,* HarperCollins, New York, 2009, p. 49.

22. Ibid, pp. 110, 128.

23. Asaf Degani and Earl L. Wiener, "On the Design of Flight-Deck Procedures," NASA Contractor Report 177642, NASA Ames Research Center, Moffett Field, Calif., June 1994.

24. Henry Petroski, *Harvard Business Review,* 82(10), October 2004.

25. The ISO 9000 family addresses various aspects of quality management and contains some of the International Organization of Standardization's best-known standards. The standards provide guidance and tools for companies and organizations that want to ensure that their products and services consistently meet customer's requirements, and that quality is consistently improved.

26. Lieutenant Colonel Greg McCleary, in an interview with the authors in July 2012.

27. Ibid.

28. Lieutenant Colonel Joe Brophy, in an interview with the authors in July 2012.

29. Degani and Wiener, op. cit., p. 53.

The Nine Principles of Automation Airmanship, Condensed

The First Principle of Automation Airmanship: Planning

Mandatory preflight briefings conducted by the PIC (or Captain, Mission Commander, Aircraft Commander) include the role of automation in the successful outcome of the flight or mission, specifically: individual proficiency, readiness, and competency; the status of aircraft systems and the impact on automated subsystems; consideration of operational conditions (weather, routing, ATC, etc.) and how aircraft automated systems can be used to manage anticipated conditions and likely contingencies.

Skills and Practices Based on the First Principle:

In order for mission preparation to succeed in creating an environment that allows the crew to perform at an optimum level, the captain, pilot-in-command (PIC), or aircraft commander (AC) must conduct a complete crew briefing, including an assessment of each crewmember's proficiency, readiness, and competency for the mission. Since most automated aircraft have specific crew functions related to the automation, from aircraft power-up to shutdown (and sometimes prior to start-up), the PIC must address critical functions during mission preparation that could impact the role of the automation during the mission. Additionally, degradation to automated systems often affects mission outcomes or even overall mission readiness, and must be addressed in mission planning. Automation duties that impact the operational mission are briefed, and additional/augmented crewmembers are issued specific instructions for critical phases of flight to include backup and monitoring duties, when applicable.

The Second Principle of Automation Airmanship: Briefing and Debriefing

Crews leverage opportunities to brief and debrief critical flight path information, automation configuration, and individual responsibilities prior to commencing maneuvers; briefings reference inputs made to the FMS and are rigorously checked against published charts while timely debriefings include unexpected/unanticipated automation behavior, incorrect setup, and individual performance.

Skills and Practices Based on the Second Principle:

To manage the quantity of information available on the advanced flight deck, crew briefings on advanced aircraft are specific and detailed with respect to automated systems (for example, when the pilot flying (PF) intends to engage the autoflight and how the autoflight will be configured for an instrument approach). Briefings should be inclusive of critical operational information, but not exhaustive in every detail: effective briefings do not contain "too much" information and thereby "weigh down" crewmembers with unimportant information and/or data. Routes, standard instrument departures (SIDs)/standard arrivals (STARs), and approaches are briefed with reference to the FMS or equivalent system, if available, and compared to map or other plan views. Autoflight mode configurations for critical phases of flight are briefed. Additionally, high-performing crews conduct timely debriefings (which can be accomplished during low-workload phases of flight), which include unexpected/unanticipated automation behavior and incorrect setup whenever they occur during any flight. Experienced, high-performing crews seize upon opportunities to learn from unexpected or unanticipated occurrences on the flight deck, by addressing safety, standards, unanswered questions, and improvement opportunities with accurate, timely, and, if necessary, detailed debriefings.

The Third Principle of Automation Airmanship: Data Entry

Data entry errors are eliminated by crews who demonstrate a high level of proficiency in finding correct input pages (or other input pathways) and entering data efficiently and errorfree during all phases of flight; crews cross-check all flight-critical data without exception and communicate input that is conducted independently, resulting in information displays that are accurate, relevant, and understood by each crewmember.

Skills and Practices Based on the Third Principle:

Automated aircraft require a great deal of expertise so that crews will enter data and parameters into the FMS and other mission systems with a minimal chance of error. Detailed and specific crew procedures, standard operating procedures (SOPs), and standard practices decrease the likelihood of error-free entries, yet aircrew must execute these instructions under high-workload, emergency, and time-compressed conditions. The highest-performing crews demonstrate ease in finding correct input pages and entering data efficiently, without error, under all flight conditions. When errors are committed, they are immediately caught, trapped, and corrected. All flight-critical data (altitudes, speeds, lateral and vertical navigation data, altimeter settings) are cross-checked by both pilots. Since automated aircraft have multiple interface units through which data may be entered by several crewmembers without direct oversight of others, it is essential that crewmembers communicate data entry that affects the current or future flight path of the aircraft. Therefore, the highest-performing crews do not engage in "secret typing," thus preventing errors from entering into automated systems. Similarly, flight and mission data displays have proliferated in the cockpits of automated aircraft, creating the need for crews to prioritize and manage information displays to reflect the most important flight and mission-related information for every phase of flight. The highest-performing crews maintain displays that are up to date and relevant to the phase of flight; "good housekeeping" is adhered to by limiting unnecessary and irrelevant information. Because programming and data input during critical phases of flight (including taxi and ground operations) effectively reduce the number of crewmembers who are looking inside or outside of the aircraft for potential hazards and threats, it is essential when crewmembers are making data inputs, that this be both communicated to the rest of the crew as well as minimized: in the most effective crews, "head-down" time is minimized during both routine and nonroutine operations.

The Fourth Principle of Automation Airmanship: Communicating

Accurate, clear, and concise communications among crewmembers are as vital as the communication (via data input) with aircraft systems; modifications to the flight guidance that impact the flight path (current, future, or contingency) are briefed, and plans are compared to objectives to ensure that the mental model of the flight path desired by the crew is consistent with input to the automation; mode system failures and degradations are annunciated by crewmembers as they occur; and challenges to automation configuration and setup are made by each crewmember without prejudice.

Skills and Practices Based on the Fourth Principle:

With the reduced crew complement and increased performance associated with advanced aircraft, CRM and airmanship skills related to communications must be raised to a higher level of effectiveness in order to avoid breakdowns which can rapidly erode situational awareness. Crews must interact with the automation within strict data entry and verification protocols established by procedures. Inputs, modifications, and changes to the automated flight path are briefed between crewmembers in order to maintain an up-to-date shared mental model of the aircraft's flight path. Crews must constantly compare their intentions for the automation with the actual performance, and they must be sure of "what it's doing now" and "what it's going to do next." Crews must have a sound understanding of priority one, priority two, and priority three messages generated by the aircraft and must use the information to plan appropriately. Smooth and effective crew coordination arises from a shared knowledge across the entire crew of task and workflow structures and the processes employed to execute flight and mission tasks. All crewmembers, regardless of position within the hierarchy of command, remain actively engaged in building, maintaining, and revising, as necessary, the shared mental model of the task at hand. Feedback and self-diagnosis are an ongoing and constant part of communications when outcomes do not match objectives or plans require modification. Pilots and crewmembers include "intent" when discussing plans and outcomes with one another, especially those related to autopilot, autothrottle, and FD/FMS mode changes. Updates to automation modes and system failures and degradations are annunciated when made by any crewmember; challenges to automation configuration and setup are met with respect by all crewmembers; and ambiguity is reduced through clear communication of intent. Communications among crewmembers that are ambiguous and/or confusing are challenged and resolved without prejudice, prior to their impact on the aircraft flight path vector.

The Fifth Principle of Automation Airmanship: Monitoring

Every crewmember must constantly apply monitoring strategies proportionate to the phase of flight, criticality of the situation, and SOPs of the operation. All monitoring must consist of an adequate understanding of both the human physiology of monitoring and the automated monitoring and warning systems on the aircraft. Crewmembers must never lose touch with monitoring that is delegated to another crewmember or an automated system.

Skills and Practices Based on the Fifth Principle:

Automated aircraft provide literally hundreds of backup functions, alerts, cautions, and warnings—aural, visual, and haptic—which allow aircrews to assess the status of the aircraft flight path and systems more easily and quickly than in less complex, less automated aircraft. This has proved to create the potential for crewmembers' vigilance to atrophy, giving the automated systems greater authority than would otherwise be acceptable. Additionally, the relative reliability and seemingly "infallible" nature of solid-state, automated systems can lead to a false impression that the automation is "never wrong." To prevent this from occurring, the flight crew must maintain clear authority over the automation, keeping the pilots and crew in proper relation to the automated system. Within the most effective crews, the "attention spotlight" is used to maximize the information that is monitored, according to a strict priority (flight path safety, clearance compliance, contingencies). When monitoring is conducted as a flight crew discipline, the automation acts with crew consent only. Similarly, highly automated aircraft are often designed to anticipate flight path changes and to make these changes automatically. Therefore, as automation suggests configuration and/or mode changes, high-performing crewmembers demonstrate full awareness and control over the aircraft. The increase in both complexity and information available on the automated flight deck requires that all crewmembers share more equally in the generated workload. Expert monitors are aware of their own cognitive limits and do not overtask themselves or members of their crew: experts brief monitoring responsibilities, "monitor the monitoring" and, to the extent possible, manage the rate of information flow onto the flight deck. Expert monitors maintain a monitoring protocol that ensures that when automated flight reverts to manual flight, sudden and unpredicted control inputs are not required, and normal manual flight is "as smooth" as normal automated flight.

The Sixth Principle of Automation Airmanship: Situational and Mode Awareness (SMA)

Modes for autoflight, flight guidance, and sensors are learned, understood, and practiced so that they support an accurate mental model of the aircraft flight path while displays reflect crew intentions during all phases of flight; crews understand the monitoring-workload equation for all phases of flight and resolve and debrief automation surprises.

Skills and Practices Based on the Sixth Principle:

Situational and Mode Awareness (SMA) merges the commonly accepted concepts embodied in Situation Awareness and Mode Awareness. These combined concepts comprise SMA, which can be defined as "the accurate, useful mental model of relevant aircraft automated tasks, including configuration, flight and powerplant states, flight guidance, flight control and sensor modes, and their dynamic relationship to the present and future flight path of the aircraft." Modes for autoflight, flight guidance, and system sensors are learned, understood, and practiced so that crewmembers can form accurate mental models that increase SMA. Information management systems and displays have been optimized by designers to be capable of displaying for the crew much more information than in traditional, legacy-type cockpits. Therefore, to maximize the crew's SMA, data displays must reflect crew intentions during all phases of flight. Cockpit displays are up to date, uncluttered, and relevant to the current phase of flight. The crew uses available displays effectively in order to not become confused or fixated with the automation. The most effective crews demonstrate understanding of the relationship between monitoring and the level of automation selected (as the automation increases, physical workload decreases and monitoring requirements increase; as automation decreases, physical workload increases and monitoring becomes less demanding). Within the highest-performing crews demonstrating high levels of SMA, automation surprises are rare, and when they occur, they are immediately resolved and debriefed.

The Seventh Principle of Automation Airmanship: Workload Management

Crews maintain a supervised, effective level of automation in order to maintain and protect the integrity of the aircraft flight path through an understanding of the relationship between manual and automated aircraft control; thus automatic and crew-assisted transitions between levels of automation occur without surprise or abrupt changes to the flight path, and mode transitions do not occur without pilot consent and awareness.

Skills and Practices Based on the Seventh Principle:

Expert crews use the capability of advanced cockpits to maximize their SMA and to optimize crew workload across every phase of flight. They balance workload and monitoring requirements to ensure that the appropriate level of automation is engaged to protect the integrity of the flight path. A significant advance in aircraft design and performance is the capability of the most modern aircraft to transition between various phases of flight (takeoff, cruise, descent, approach, etc.) automatically and without crew input, creating a smooth, stable, fuel-efficient, and predictable flight path. Most aircraft automated systems indicate these transitions to the crew with a variety of cues both aural and visual, and many can accept full or partial crew intervention in order to alter the flight path vector at the crew's discretion. In the highest-performing crews, both of these actions (automatic and crew-assisted transitions) occur without surprise and without unnecessary abrupt changes to the flight path. Within these crews, aircraft flight path and mode transitions are smooth, anticipated, and consistent with mode selection and air traffic control (ATC) clearance. Additionally, the most effective crews understand the relationship between manual and automated aircraft controls and systems, and they apply this knowledge to the transition from manual to automated flight and from automated to manual flight. Therefore, these crews make manual-to-automated flight and automated-to-manual flight transitions smoothly and without surprising other crewmembers. Likewise, "silent mode transitions" do not occur without pilot consent and/or awareness.

The Eighth Principle of Automation Airmanship: Positive Flight Path Control

Any unexpected in-flight event or failure requires that the pilots fly the aircraft first, protecting the integrity of the aircraft flight path vector prior to any other cockpit responsibility of lesser importance. Subsequent cockpit duties can be carried out only after the flight path has been stabilized and flying responsibilities have been delegated to a crewmember or the autoflight system.

Skills and Practices Based on the Eighth Principle:

Modern, highly automated aircraft are designed with multiple levels of warning systems, visual, aural, and haptic. Additional, enhanced and specialized systems are commonly added to the flight deck throughout the life cycle of many aircraft, making the knowledge of these systems a dynamic process for all aircrews. Many advanced technology aircraft have unique display and prioritization systems that help crewmembers organize and prioritize nonroutine, abnormal aircraft conditions and system failures or degradations. Expert crews have a sound working knowledge of the stress-performance relationship, and they have a deep personal knowledge of their own behavior during unexpected, stressful events. Because modern alerting systems are typically more complex than those on traditional, legacy-style aircraft, it is not uncommon for expert crews to also have a sound working knowledge of both the warning system itself and the checklist or other flight crew interfaces which are designed to assist the crew to an effective, safe resolution. Applying this knowledge with skill and expertise results in crews handling nonroutine alerts and warnings smoothly and accurately. Experts in the field of accidents and complex system failures advocate that when complex systems fail, they fail in complex ways. Automated aircraft failures are well documented, and examination of accidents and near-accidents reveals that automated aircraft are neither fail-safe nor immune to ambiguous and misleading information being presented to the crew during system failures. Therefore, crews of advanced aircraft must "fly first," in addition to being skilled in evaluating system failures, in order to create successful, safe outcomes with minimum disruption to mission accomplishment. For these crews, degradation to aircraft systems has minimal impact on flight deck organization and aircraft flight path. As automated systems fail (autopilots, autothrottles, etc), crews must be skilled in recovering the aircraft to a safe and stable flight path, as well as in establishing a safe clearance from terrain, enemy threats, weather, and other aircraft. For the most effective crews, recovery from flight path deviations and/or system failures is smooth and efficient. Pilots demonstrate ease in selecting an increase or decrease in the level of automation in response to failure. For crews to assimilate aircraft characteristics and adopt future defenses against unsafe outcomes, all failures and deviations that occur during a mission are discussed and resolved with other crewmembers with specific debriefings, during flight or immediately after the flight concludes.

The Ninth Principle of Automation Airmanship: Logic Knowledge

Through a sound understanding of the basic logic of the automated systems and procedures, crews are able to maintain a sound mental model of the aircraft state at all times; knowledge-based modeling of expected and anticipated autoflight actions is consistently accurate, which prevents crews from under- or overreliance on automation for flight path control by balancing their cockpit scan between relevant cockpit displays and, when appropriate, outside-the-cockpit clearing and scanning.

Skills and Practices Based on the Ninth Principle:

Automated systems capable of controlling the vertical and lateral flight path with tremendous precision do so by using complex algorithms and control devices, of which most crewmembers do not have, or need, a detailed working knowledge. However, the highest-performing crews have a sound understanding of the basic system functions, and they consistently develop accurate, sound mental models of the state of the aircraft to assist them in interpreting and projecting events associated with the overall automated system. High-performing flight crews apply this knowledge of their FMS, mission systems, autoflight system, and flight guidance system as well as knowledge of automated major aircraft systems to their decision-making process. Their crew communications and briefings include an accurate modeling of expected and anticipated automated actions. Situational and mode awareness is positively impacted by the crew's knowledge of flight and mission computer logic. Following traditional training practices used in most legacy communities, during the early stages of their introduction pilots of advanced aircraft require increased time to develop the same high level of confidence and proficiency that they demonstrated in previous aircraft flown. With the development of discrete, trainable, measurable skills for operating highly automated aircraft, pilots are able to accelerate the rate at which they develop expertise, and they demonstrate much higher confidence levels, and sooner, than many of their predecessors. Crews with a high level of understanding of the automated system, and therefore a high level of automation confidence, do not demonstrate overreliance on automation; their cockpit scan is balanced between HUD/HDD/navigation displays and Visual Flight Rules (VFR) clearing when necessary. Similarly, those pilots do not underrely on automation, using automated systems to effectively manage workload and to maximize situational awareness. Crewmembers demonstrate ease in using and discussing automation, including the use of correct terminology and technical language, when describing and interacting with the aircraft automated systems and other crewmembers.

Index

"f" indicates material in figures. "n" indicates material in notes. "t" indicates material in tables.